PENGUINS IN THE DESERT

PENGUINS

IN THE DESERT

ERIC WAGNER

Oregon State University Press   Corvallis

Library of Congress Cataloging-in-Publication Data

Names: Wagner, Eric Loudon, author.
Title: Penguins in the desert / Eric Wagner.
Description: Corvallis : Oregon State University Press, 2018. | Includes index.
Identifiers: LCCN 2017054219 (print) | LCCN 2018001632 (ebook) | ISBN
    9780870719271 (ebook) | ISBN 9780870719240 (original trade pbk. : alk. paper)
Subjects: LCSH: Magellanic penguin—Argentina—Tombo Point. | Tombo Point
    (Argentina)
Classification: LCC QL696.S473 (ebook) | LCC QL696.S473 W34 2018 (print) |
    DDC 598.47—dc23
LC record available at https://lccn.loc.gov/2017054219

♾ This paper meets the requirements of ANSI/NISO Z39.48-1992
(Permanence of Paper)

Photos by Eric Wagner unless otherwise credited.

Oregon State University Press
121 The Valley Library
Corvallis OR 97331-4501
541-737-3166 • fax 541-737-3170
www.osupress.oregonstate.edu

for El

Contents

Author's Note

> Of course, he worked from memory and memory is usually too inventive.
>
> Jorge Luis Borges

This book is an account of the time my wife, Eleanor (hereafter, El), and I spent among the Magellanic penguins of Punta Tombo, Argentina, in 2008 and 2009 as volunteers with the University of Washington's Penguin Project. I kept a detailed journal and have since spoken at length with the various principals, but I was no stenographer. What I mean is that conversations and events are recounted as best as I can recollect them. The views presented here are likewise my own, and not necessarily those of the Penguin Project. No one is to blame for my opinions but me.

I have included research in the broader discussions of penguin behavior and ecology that was sometimes completed or published after my time at Punta Tombo. In part, I wanted the science to be as current as possible, but I also think one of the consolations of natural history is its timelessness. Many if not most of the phenomena scientists are just now beginning to understand have been in motion long before we got around to describing them.

Finally, Turbo is real.

Penguins in the Desert

> Everywhere animals offered explanations, or more precisely, lent their name or character to a quality, which like all qualities was, in its essence, mysterious.
>
> John Berger, "Why Look at Animals?"

Think of a penguin. What do you see? If you are like I was a few years ago, you see a cathedral of ice and snow. You see towering glaciers. You see a cold, dark ocean with great floating bergs so white they seem to glow blue. Above all, you see a forbidding landscape thousands of miles from anywhere, and in the midst of its august emptiness, you see a group of monkish black-and-white birds huddled against the unrelenting winds, the very picture of animal stoicism.

Now let me offer a different scene. Instead of ice and snow, think of sand and dust and dirt. Instead of bitter cold, think of blazing heat. Instead of a colony of birds far removed from people, think of one just a couple of hours from a city, eminently reachable on any old summer afternoon. (You can keep the wind, however.)

Scientists recognize eighteen species of penguins, but only five breed in Antarctica, and just two of them solely. The rest are scattered throughout the Southern Hemisphere, on small islands in the middle of the oceans, along the coasts of South Africa and Namibia, Australia and New Zealand, and from the southernmost tip of South America all the way up to the Galápagos Islands at the equator. No other family of birds lives across such a wide range of latitudes, but while these penguins of the temperate zone far outnumber their icebound kin, they have received considerably less attention. Frankly, we are missing out.

In September 2008, my wife, El, and I traveled to a place called Punta Tombo, on the coast of Argentina. Punta Tombo is home to a colony of Magellanic penguins. With a population of somewhere between 1.1 and 1.6 million breeding pairs, the Magellanic penguin has colonies scattered along both the Pacific and Atlantic coasts of South America, out even to the Falkland Islands (or the Islas Malvinas, as Argentines call them). But their biggest gatherings are in Argentina, and Punta Tombo, having more than four hundred thousand birds, is the biggest of all.

El and I had gone at the behest of Dee Boersma, a biologist at the University of Washington. I was a graduate student studying oysters at the time, thirty and somewhat adrift, and close to leaving the program altogether when Dee saw a chance to intervene. She and I were talking one day in her office, and as I was just about to go she said, "You and your wife wouldn't want to take six months off and come measure penguins, would you?"

I stopped at the door, half in, half out. "Seriously?" I asked.

"Sure," Dee said, bouncing on the big blue yoga ball she used as a chair. "But just so you know, it's fourteen hours a day and there are no days off. You'll live in a trailer. And it's all on a volunteer basis, though we'll take care of your expenses while you're down there."

"Let me check with El," I said. I hurried to a coffee shop and dashed off an email. "What do you think?" I wrote. El was an editor then at one of Seattle's small community newspapers. It was a frantic job, and I didn't imagine I'd hear from her for a while. Between the two of us she is the planner, and we could discuss penguins in the evening, weigh the pros and cons, and so on.

She wrote back four minutes later. "Yes! Yes yes yes! I LOVE DEE!"

So that was settled. A couple of weeks later, we were in the penguin lab at the University of Washington, listening, enraptured, as Dee talked about what she said would be a once-in-a-lifetime opportunity. Her presentation was compelling, but I expected no less. I had met her a couple of years before, when I took her class on conservation biology. Her lectures were memorable for the energetic brazenness with which she could hold forth on the way things ought to be. It was inconsistent with the objectivity I thought the dictum of science, but Dee was unapologetic. Conservation biology is an applied discipline, she explained. A system of values is central—essential—to its practice. "You can't listen to everything I'm telling you and not feel anything," she had said.

Dee could get away with this because she had studied penguins for forty years and is one of the world's foremost conservation biologists; a writer for the *New York Times* once called her the Jane Goodall of penguins. The birds of Punta Tombo could accommodate her beliefs precisely because they themselves were incapable of political nuance. If the penguins became coated with oil while at sea, they washed up dead on the shore. If a changing climate scattered the fish they relied on for food, they swam until they found them or starved. The colony had been declining for decades, and just four years ago the International Union for the Conservation of Nature had added the species as a whole to its Red List as "Near Threatened." In the face of such circumstances, it was up to the scientists—Dee and now, perhaps, El and me—to watch the penguins, learn from them, and act as conscience dictated.

Of course we agreed to go, but even in the throes of our initial enthusiasm, I had nagging doubts. I wondered how we would handle the rigors of Punta Tombo. I had done field work for a few summers before, but this would be El's first prolonged exposure to The Field, and I knew from experience that The Field can be a challenging psychic landscape. The Field is either frigid or sweltering, sodden or parched, but never in between. That hordes of biting flies patrol The Field is taken as read, but you must also vie for your bodily fluids with fleas, ticks, and leeches. Your clothes are never clean in The Field and the food is utilitarian at best. Don't even think of drinking the water.

Those are just the physical privations. Punta Tombo would present unique challenges. "A lot of penguin chicks are going to die," I warned El one day while we were packing. She liked science well enough and was comfortable with spare living, but she'd studied literature and linguistics, not biology. More to the point, she is to her core a staunch champion of vulnerable creatures.

"That's okay," she said. "I know how nature works."

"They might die horribly," I went on. "Right in front of you. And you couldn't do a thing about it."

"Mm," she said.

I peeped pitifully and fluttered my hands the way I thought a dying chick might.

"Stop it," she said.

This, then, would be the last and most serious consideration: to work in The Field as a scientist is to assume a particular relationship with a study subject. We would observe the penguins, we would become more familiar with their habits than we were with those of our immediate families, but we would do so from a distance, in the manner of accountants.

We discussed these provisos, assured ourselves that we were sufficiently, spiritually robust, and then off we went to Punta Tombo, to The Field—to the place where, as the ecologists like to say, we would confront our models with data.

1

In Patagonia

The first penguin stands in the arrival hall at the Almirante Marcos A. Zar Airport, just outside the city of Trelew, in southern Argentina. It is six feet tall and made of plaster. A large chick lies at its feet, but the penguin has eyes only for the deplaning passengers, at whom it stares with an unblinking gaze.

The next penguins are amassed on the highway outside the airport, on a billboard that promises *1.000.000 de Pingüinos de Punta Tombo!*—a promise that has not been true for more than thirty years. In Trelew itself, penguins—pair after pair of them, nuzzling one another—are stenciled on the taxis parked around the central plaza. A few blocks away, a penguin is emblazoned on the placard of the Hotel Centenario, one of the city's oldest tourist lodgings. More penguins are for sale in stores or from street vendors, as plush dolls, postcards, stickers, key chains, other bits of kitsch.

The next penguin is not a penguin at all, but a white pickup truck called Opus, after the penguin from the *Bloom County* comic strip. Having gathered three passengers, Opus sails out of Trelew and down Ruta Nacional 3, the highway that runs the length of the Argentine coast. After a few miles, the truck leaves the highway in favor of Ruta 1. A provincial road, Ruta 1 is emptier, lonelier, and also unpaved. Opus rumbles along, its tires kicking up stones, and the city disappears behind in billows of dust, while the worn hills of the Patagonian steppe spread out under the pale blue sky.

The next penguins are forty-five miles or so down Ruta 1, on a sign that points the way to the Área Natural Protegida Punta Tombo. Here,

Opus turns left and thumps over a sheep guard. This new road is narrow and treacherous, with corners of trigonometric daring. Around these Opus flies, flinging stones. The truck bears down on the wildlife, sends them all scurrying into the bushes: herds of guanacos, a relative of the llama; lanky European hares; and elegant-crested tinamous, small chickenlike birds that fly only as a last resort. Ten miles later, the truck sways around a bend, and there in the distance is the point itself, Tombo, a gnarled finger of arid earth that reaches into the Atlantic Ocean, a blue so rich and cool it seems out of place.

The road comes to an end in front of a squat, beige house made of cinderblocks. Behind the house is a Quonset hut, also beige, and two old camper trailers. They were not beige when they were new but are well on their way to becoming so now, after many years in the sun and wind.

The first penguin—the first real, live Magellanic penguin—is just past the house and the hut and the trailers. It lies on its belly under a big scraggly bush next to the gravel tourist trail, its flippers tucked against its sides. It appears to be dozing, but at the crunch of human steps it opens first its right eye, then its left. It cocks its head and appraises the newcomers who stand before it: me, El, and Ginger Rebstock, a scientist from Dee's lab. We four consider each other for a moment. Then the penguin closes its eyes and seems to forget we are there.

Ginger, El, and I continue along the trail until we reach the crest of a small hill. On the hill's seaward side, the landscape is transformed. Penguins are everywhere, hundreds of them, thousands, maybe tens of thousands. Some wander among the scrubby bushes, either singly or in packs of four or five, like gangs of diminutive street toughs. More bask in front of the bushes, or relax in the mouths of burrows. The early evening sun lights their white bellies and black faces. Aside from the steady throb of the Atlantic, the scene is still, calm. As a collection of animals, and famously raucous animals at that, these penguins are almost eerily peaceful.

Then one bird lying near us gets slowly to his feet, fluffs out his feathers, and lifts his head.

⌒

The Magellanic penguin is, at a little over two feet tall, medium-sized as penguins go. Males are larger than females, but the differences are subtle, down to the thickness of the bill a few millimeters this way or that, or the squareness of the head. The body is stout but streamlined, from the head to the compact neck and the round, sturdy back. The flippers hang to the side like afterthoughts. Although the species has a generic penguin's black back and white belly, to this template it has added its own subtleties of dress. The matte black of its back branches at the neck to form a collar around the throat before continuing to the top of the head. A separate band, curved in an inverted U, splashes across the chest and dribbles down the flanks before spotting off at the tail. A black face mask covers the cheeks and chin. The eyes are subtle and complicated, and range in color from pink to red to brown, depending on the bird's age. The pupils have a colored ring around them as well, which can be either pink, red, or gray.

"An animal's eyes have the power to speak a great language," the philosopher Martin Buber has said, but do not forget the animal's voice. In addition to the Magellanic penguin, three species—the Humboldt, the African, and the Galápagos—belong to the genus *Spheniscus*. Before each received its specific name, they were known in loose collective as jackass penguins, since their territorial call, or ecstatic display, can sound uncannily like a donkey's bray. A physically demanding spectacle, the Magellanic penguin's ecstatic display starts when a male stands in front of his nest and feels for whatever reason a pressing need to assert his claim. He warms up with a series of huffs—*hu . . . hu . . . hu*—before extending to his full height. His flippers flung wide, he points his face to the heavens, his bill agape, his breast and belly heaving like a bellows: *Hu-hu-huhu-hooAAAAAAAAAAAH. Hu-hu-hu.* Having so concluded, he waits to see if another male dares answer his challenge. If one does, he brays again. If the rival is especially impudent, the two might bill-duel, clacking their bills rapidly against one another. If all his neighbors are sufficiently cowed, the male settles back on his heels and puffs out his chest, a picture of contented virility.

A Magellanic penguin brays
outside his burrow.

We watch now as this male in front of us brays. Another male nearby picks up the call. A little farther off, other males start to bray, sometimes for the benefit of a neighbor, sometimes for purposes less immediately clear. Their hoarse calls cascade over us, and I am reminded of an orchestra warming up before a concert, as each player prepares to perform some great work written ages ago. For these penguins, this is a final rehearsal of necessary talents.

Ginger, El, and I turn to leave. Tomorrow we will make our more formal introductions, but now it is late, and we are cold and tired. We trudge back to the house as the sun sinks beneath the low hills in the west, and the sea pounds out its rhythms, and, all around us, the penguins wail against the growing dark.

⌐◦

The next morning, I wake early and poorly rested. This is due in part to the jet lag—Punta Tombo is five hours ahead of Seattle—but the penguin braying right outside the trailer most of the night didn't help. At times, he sounded like he was directly under my pillow, his bill aimed at my ear. Once dressed, I stumble out to find my tormenter resting quietly in the trailer's undercarriage: a penguin in repose. I glower at him. He looks distinctly untroubled. Jackass penguins indeed.

El comes out a couple of minutes later—she says she slept fine—and we head to the house. It is not yet seven o'clock, but Ginger is bustling about. For the next few weeks she will be our guide to this place. She is in her early fifties, short and trim in appearance, with dark, solemn eyes and a soft, serious face. She has a PhD in biological oceanography and started working for Dee after several years of doing research on small crustaceans called copepods. She is friendly, but also exacting and precise, and can at times be a tad officious. Sometimes when I ask a question, she has a way of looking at me, ever so slightly askance with a fretful purse of her lips, which can make me wonder whether I absolutely needed to open my mouth. Before she left Seattle, she had her already short hair trimmed to a fine gray fuzz; I wondered if she was surprised El and I had not done the same. Still, being in the field has brought out an impish side I didn't know she had. On the drive down from Trelew, she talked of how she liked to take the choruses from songs and fit penguin-related lyrics to them where metrically appropriate. She sang us one example, from "Home on the Range":

Oh, give me a hoooome where the *pingüinos* rooooam,
And the cuis and the guanacos plaaay

Ginger, we learn, has already collected some data, having gone next door to visit with the Argentine *guardafaunas*, or park rangers, who occupy the house's other half. The head *guardafauna*, a friendly, balding fellow named Miguel, arrived at Punta Tombo near the end of August. He told Ginger the first penguin came ashore soon after, on August 29. A couple of days later he found its mangled body. He thinks a *colpeo* fox killed it.

"Probably not the best start to the year," Ginger says.

After breakfast, she leads us to the Quonset hut, which in the argot of Punta Tombo is called the *cueva*, or cave. Inside it is dark and gloomy and musty, with slumping stacks of cardboard boxes and plastic tubs. If an organizational scheme exists it is not obvious to me, but Ginger starts to pick her way through the ecological hoard. "Okay," she says. "Let's get you two set up."

Before El and I left Seattle, it was impressed upon us that the Magellanic penguins of Punta Tombo have an extended reach. Although the details of their story are local, the themes are immense. Through them we will see the power of climate change, the health of the oceans and the life therein, the natural history of this singular landscape. Nevertheless, before we can tease out the ways climate change will affect not only the penguins, but also the broader ecological webs in which they are situated; before we can look for links between penguins and the state of the fish and the seas; before we can explore the ways humans are or are not willing to share the planet with its nonhuman denizens, we must first dress ourselves in the uniform of Punta Tombo. This includes:

One green or brown shirt: The *guardafaunas* insist we dress in earth tones to match the landscape, so as not to detract from the tourist experience by too visibly flaunting our access to the colony's otherwise forbidden areas.

One pair of green or brown pants: See above.

One pair of plastic kneepads: The ground is hard and pebbled and we will often be dropping to our knees. Protection is essential.

One pair of gaiters: To deter the fleas that swarm the nests. Like most penguins, the Magellanic has its own species of attendant flea. The flea yokes its life cycle to the penguins, but should a few biologists arrive first, we are more than acceptable as substitutes.

One fishing vest: Preferably with plenty of pockets, so there is room for:

Five stainless steel flipper bands: Perhaps the most important item, bands are used to identify the study penguins, which wear them on their left flipper. Each band has a five-digit number. Those we carry this year start in the sixty-thousands, meaning that about sixty thousand penguins have been banded since Dee began working here in 1982. We will band more than three thousand adults and chicks this season.

Five yellow cattle ear tags: For attaching to any bush or burrow in which a study penguin has decided to nest.

One roll each of pink, green, blue, and orange plastic flagging tape: For marking cattle-tagged nests that study penguins use during the season, so a glance can reveal the nest type. Study penguins belong to one of four categories: known-age, not known-age, satellite, or priority. Known-age penguins (KA) are those whose hatch-year and possibly even hatch-day is known. Their nests get a strip of green tape when the study bird arrives, and strips of blue and pink tape when its mate returns. Not known-age birds (ØKA) are those banded as adults. They get a strip of orange tape. Satellite birds, or sat-valuable birds (SVB), have worn a satellite tag at any point since 1995. Their nests get strips of orange and green tape. Priority birds (PR) were banded in 1982 and 1983, during the project's first season. Dee banded many chicks, but also a fair number of adults, meaning that since most penguins do not breed until their fourth or fifth year, these penguins are over thirty years of age. Their nests get blue and orange tape.

One spring scale (6 kg) / One dog's leash: For weighing penguins. The leash is looped under the flippers and hooked to the scale. The penguin is then hoisted aloft and its weight recorded.

One set of calipers and a ruler: To measure various penguins' various parts.

One band book: A small laminated spiral notebook that contains all the band numbers. We are to consult the band book whenever we come across any banded penguin anywhere, so we can tell at a glance the year it was banded, whether it is a known-age bird, whether it has been measured recently, or whether it has ever been seen since it was banded.

One waterproof notebook: So we can record every band we see as we walk around the campo, and also so we can note our stray thoughts, stray data, natural oddities, unnatural oddities, etc.

One compass: For finding our way.

One map of the entire colony, and a sheaf of individual area maps: As part of her study, Dee has divided the colony into several areas. Over the years there have been more than twenty, and of those, twelve are still active. The maps show the area boundaries, along with the sometimes very approximate locations of the study nests. Ginger says we will spend most of our days now stumping around the colony, map in hand, while in the other hand we carry:

One gancho: A piece of rebar about three feet long, the last few inches of which have been bent into a shepherd's crook and liberally wrapped with duct tape. (*Gancho* is Spanish for "hook.") We will use the *gancho* to hold a penguin by its foot so we can handle it.

All of which we are to apply to the study of:

One penguin: Spheniscus magellanicus.

꙰

So it is that we set out. Our first work of the day, of the season, is at an area near a *cañada*, or dry riverbed. Ginger explains we will check the area, called, appropriately enough, Cañada, first thing in the morning for the entire season. It will be the axis around which our world revolves, as we track all the penguins nesting within its boundaries as closely as is scientifically feasible.

With that, she strikes off across the campo. Her pace is brisk, her gaze fixed straight ahead. El and I hustle to keep up. The morning is chilly, the sky a flat gray. Plants grab at me with their long thorns, insisting I learn their names.

Ginger reaches a weathered wooden stake jammed in the ground next to a large *Lycium* bush. From her backpack, she pulls a yellow binder with "CAÑADA" written across the cover in block letters. She opens to the first page and summons lead to her mechanical pencil with a crisp *tic tic*. "All right," she says. "Let's get started."

At the edge of the bush is a depression in the dirt the size of a large fruit bowl. An old yellow cattle tag hangs next to it, lashed to a twig with a piece of green baling wire. The tag says "822I" in faded black marker. This is the first nest.

"Is there a penguin in it?" Ginger asks.

El and I look at the depression. Some sticks are strewn about it, as well as the odd pebble, but we see no penguin, and say as much.

"Okay," Ginger says. "So we put a zero and the time." She scribbles, flips the page. "The next nest is 822K, one meter away, north-by-northeast."

I turn on my heel and find this new nest, 822K. No penguin in that one, either. Nor are there penguins in 822F, 822D, 817E (although one is loitering about five meters away, which Ginger makes a note of), 817C, or 817H.

After last night, when Punta Tombo looked overrun with penguins, it is odd to document how few are actually here. Ginger tells us the colony is not even half full. At this point, only the most aspirational males have returned. These tend to be younger birds, perhaps breeding for the first time. They wager that if they arrive early enough they can claim the best nests for themselves. It seems a sound strategy, but soon the older males will arrive to oust the younger birds. In so doing, they will teach a central tenet of penguin natural history: males often claim the same nest year after year after year.

We continue our circuit of penguin-less nests. It is not until the twelfth nest, C02A, that I see a male penguin resting next to the cattle tag. When we approach, he regards us the way one might a trio of meter readers who keep bumping into each other.

"This nest has a penguin," I say, although I suspect Ginger can see that.

"Great," Ginger says. "Does he have a band?"

I check the bird's left flipper. Glinting near its base is the gray steel of a band. It is strangely exhilarating to read out the number: 53164.

"Fine," Ginger says. "So we write the band number and the time in the book." She does so. As we move on to the next nests, I feel a thrill at having completed another of Punta Tombo's many rites: first our arrival, then the first penguin, then the clothes and gear, then the first nest, and now the first nest with a banded penguin. Who can say what the future will bring?

"The next nest is C23C, southeast, twelve meters away," Ginger says.

El and I pace off twelve meters, and in what seems like the right place is a big bush festooned with cattle tags and shreds of bleached ribbon. I see tags for C23, C23A, C23B, C23D, C23s forever, but no C23C. We go around the C23 bush once, twice, a third time, and, just to be safe, the adjacent bush. No sign of C23C. I move branches to look deeper into the bush and get scratched for my trouble.

"Do you see it?" I ask El.

She shrugs. "Nope."

The back of my neck starts to heat up. Twelve meters is the longest distance we've had to walk between nests, but it is still only a little less than forty feet. I have walked forty feet many times in my life and not thought twice about it, but now forty feet feels like the chasm between the known world and chaos; or, nearer the point, the chasm between my self-perceived competence here and my actual competence.

Throughout Ginger observes us, silent and expressionless save for the occasional amused twitch to her lips.

I walk back to the last nest, pull out my compass, and pinpoint the exact trajectory the southeast line follows. I pace out what I'm sure is 39.37 feet and, rooted to the spot, stare at the C23 bush with such ferocity that I half expect it to burst into flame. Then I see, in a small bush next to the big one, the C23C cattle tag. Somehow it was tilted in such a way as to be invisible from all but this angle.

"I found it," I say, snorting through my nose. This is no triumph.

"Good for you," Ginger says. "And is there a penguin in it?"

"No," I say. "No penguin."

⌣

This routine of general disorientation fills our first days. We wake at dawn, note the weather, check Cañada, and spend hours searching for nests at sites all over the colony. One day we are in Factura, a massive bushy area so called because its lumpy outline recalls an Argentine pastry of the same name. The next day is Doughnut, a vaguely circular area that envelops the Cañada (the hole, as it were), traverses a variety of habitats, and never seems to end. There is Bungalows, a smaller area on a hillside with burrows stacked one atop another. We pair it with Sea-Tip, a wedge-shaped patch of land bracketed by the sea on one side and the tourist trail on the other. Tourist Trail adheres to the tourist trail, then extends into what Dee calls the Zona Mas Densa, which has the densest collection of burrows in the colony. Hill-Draw sends us clambering up and down hills and draws, and so on. (Naming being the act by which we confer meaning, I think it's notable Dee has named several areas after baked goods. Even Bungalows goes by a more edible shorthand, Bun.)

Through all this the penguins trickle in almost imperceptibly, until nearly every burrow and bush has one in it, and sometimes two or three. More banded birds have joined the general throngs, and Ginger tells us

The area known as Zona Mas Densa, with the densest concentration of burrows in the colony. (Photo by Dee Boersma)

some of these will have to be weighed and measured. We therefore face another Tombo rite: the penguin wrangle.

I am standing in front of 301X, a nest in Sea-Tip. The male penguin glowering out at me, band number 35472, was banded as a chick in 1989. He has nested under this small bush for five years and was first measured as an adult in 1995. After thirteen years, it is time to take his vitals again.

"Go ahead and get him," Ginger says.

I look at 35472. He wags his head back and forth, looking at me first with his right eye, then his left. Called the head-wag, the gesture is one of aggression. *Back off*, 35472 is advising, *or there's going to be trouble.*

A male penguin in the midst of an intense head-wag.

"I won't hurt him?" I ask, hoping to make clear the depth of my concern for 35472's well-being.

"No," Ginger says. "Penguins are incredibly strong. There's no way you can hurt them. Just make sure you get a good grip on his neck." She shares a few gruesome anecdotes about people who didn't get good grips. Slashed-up arms and hands figure prominently, as do pierced belly buttons.

During this recital 35472 begins to suspect something is afoot. He puffs himself up and starts huffing, head-wagging with an equally ratcheted intensity. I reach into his nest with my *gancho*, hook him by his stubby leg, and haul him out tail-first. He drags his bill in the dust, a surprisingly affecting act of resistance. I crouch over him, my hand poised like a snake to grab his neck just as Ginger taught, but when theory meets practice something goes awry. I snatch at him and miss, he expresses his vehement disapproval, and we scuffle a bit. A one-handed grip no longer seems sufficient, and I end up hefting him up by the throat with both hands. Throttling him, in other words.

"Maybe don't hold him like that," Ginger says, wincing. "I mean, he's fine, but a tourist might see you and it wouldn't look good."

A fair point, but if I didn't think it would be sporting to use my advantages in height and weight before, I have no such hesitations now. I hurl 35472 to the ground and sit on him. He struggles and thrashes beneath me, and as I wrestle him into position, I realize there are certain things you cannot know about the Magellanic penguin, cannot understand, until you are trying to restrain one between your thighs. Only then do some of the penguin's less obvious physiological attributes come into rather horrifying focus. For example: A penguin's chest all but bursts with thick slabs of muscle, and the flippers with which 35472 is spiritedly beating me are solid bone. His bill, which is clacking away perilously close to my fingers, is heavy, black, ridged, and very sharp. The upper mandible ends in a hook that slots neatly into the lower mandible and has been honed by evolution's careful craft to grip fish or squid, or, if those aren't around, my pants. The red of his eyes, too, can look surprisingly demonic.

Once I have secured 35472, Ginger measures the length and depth of his bill and the length of his flipper and foot. 35472 stares up at me balefully the entire time. When Ginger has finished, we slip the leash under his flippers and hoist him with the spring scale. It takes a while for him to stop kicking, but when he does, we learn he weighs 4.7 kilograms, which is more than ten pounds. Then I grab his neck again and lower him to the ground and let him go. He rockets back into his nest, turns, and brays in my face. I can see past his barbed tongue, all the way down his pink throat lined with fleshy protuberances called denticles. I can smell his warm breath. I can feel the full force of his fury and terror.

Emily and Ginger show the proper way to wrangle a penguin.

Over the next few days we get a lot of practice in the art of penguin wrangling. The penguins don't like it, obviously, but for my part the intimacy a wrangle affords quickly puts to rest certain notions of penguinness I had brought with me to Punta Tombo. If I once saw in a penguin's waddling mien an amusing caricature of human self-regard, then I now know I must reconcile that with other, less savory behaviors. Penguins have no qualms about using the flattened remains of long dead offspring as nesting material, for instance. They are also committed solipsists and can show a breathtaking indifference to the welfare of their neighbors. Their subjectivity swings mostly between two states, what we call Me and Not Me. These are about what they sound like. When I grabbed 35472 and he vigorously objected, we would say he was deeply in the Me. The penguin lying not two feet away that couldn't be bothered to open its eyes? Not Me.

Speaking as one who has spent far too much time stewing over real and imagined slights, I envy the penguins their self-possession. To be so fully present, to distinguish so quickly and coherently between a legitimate assault and everything else, is, for me, the essence of animal being. These are not the raccoons begging for dog food outside my house, or the crows picking through my garbage. These are affectless penguins, and this is the desert. Whenever I accost one, I hope it recognizes how much I appreciate the distinction. I hope it sees me, if not as a friend and ally, then at least as a fellow aesthete. Forcibly subduing it, clasping its neck while pinning its flippers against its body with my legs, I murmur my admiration into the feathers around its aural cavity. None of the penguins ever seem to understand. Instead, to a penguin, they glare back with the purest expressions of fear and hatred I have ever seen.

All save one.

⤳

Near the end of the third day, we are huddled around the table filling out notebooks and trying to stay warm when I hear a sharp *rap-rap-rap* at the door. El and I look at each other, confused. It is late. Through the house's thin walls we can hear the *guardafaunas* watching a soccer game next door. Who could it be at this hour?

"Better go see who it is," Ginger says.

I open the door. A male penguin is on our concrete stoop, clearly expecting to be let in. He toddles to the middle of the room, turns, and cocks his head at me. I stare at him. Ginger watches all of this from her chair. "And now you can say you have met the famous Turbo," she says, and smiles.

Early in the breeding season of 2005, as the story goes, a young male penguin was brutally evicted from his nest by an older, stronger male—hardly a rare event at Punta Tombo. Perhaps to compensate for his shortcomings, as young males of many species are wont to do, this newly homeless penguin moved under Opus, our giant Ford F150 Turbo pickup. His choice was biologically defensible. Opus provided good cover, and its chassis was nice and snug. To this penguin's discerning eye, it was a fine nest, save for its occasional tendency to drive away. Before long this penguin took to visiting the field house. He was not at all shy around humans, and so, being an oddity worth tracking, the field crew gave him a band (53080) and a name: Turbo.

Turbo stayed under Opus for the rest of that year, but the next season he moved to the center of a large *molle* bush near the house. He lives there now, his nest surrounded by coils of discarded barbed wire. But even after claiming one of the most secure nests in the entire colony, he still likes to visit the house. He seems to consider it his second home and treats it as a penguin does its own nest: he brays outside the door, clucks while in the kitchen, and sometimes scoots under one of the bunks in the back room to doze.

Turbo is at our door again the next night, and the next. His visits are soon a regular part of our evening. He comes to the door and raps on it with his bill. When we open it, he waddles in and stands in the middle of the room waiting for someone to stick out an arm or a foot for him to flipper-pat. (Flipper-patting is a precursor to copulation. Basically, Turbo is trying to mate with us.) Once he has spent his affections, he rests under the table or next to a chair while we return to whatever it was we were doing. After some interval, when he has had enough of our company, he stands by the door until we let him out. Yet he never simply leaves. He likes someone to walk him to his bush and will wait patiently on the doorstep until we oblige him. Then, checking to make sure he is accompanied, he toddles home.

Turbo stands outside the *cueva* door. (Photo by Dee Boersma)

For us, Turbo becomes a spiritual balm. Four hundred thousand penguins breed at Punta Tombo, or try to. Of those, 399,999 may fear us, flee from us, snap at us, or beat us with their flippers, but Turbo runs out to greet us whenever we pass his bush. Where the other penguins deal in Me and Not Me, Turbo is somehow, indubitably, an I. The funny thing is that he seems otherwise secure in his penguin-ness. It isn't that he thinks he is a human. Rather, he thinks we are penguins. The way we see ourselves in them, he sees himself in us. With this breach, he exposes the artifice of our separations. So released, El and I indulge in wild fits of anthropomorphism. All the things we wish we could do with all the penguins but do not dare, Turbo lets us do to him. We coo over him and caress the firm pelt of his feathers. We scratch the back of his neck as he closes his eyes in pleasure. We even pick him up and carry him short distances before he nibbles on our arms to remind us to put him down.

Such are the metaphysics of Turbo. With the newness of Punta Tombo, to say nothing of the long days so far, El and I are often completely drained. While we are thrilled to be in this spectacular place, there is so much to learn, so much to keep track of, all in the service of a single dreary fact, that the colony has declined by about 40 percent since 1987. The decline is due to a host of reasons—oil pollution, overfishing, climate change, others—but set against the hordes of penguins I daily walk past, these reasons can blur into abstraction. Then I think of Turbo sitting in his bush all alone, surrounded by barbed wire. I imagine him going to sea and having to navigate through a labyrinth of fishnets as he swims hundreds of miles and finds nothing to eat. I imagine him coming home and, on his way to his bush, getting run over by a tour bus. Through these imaginings, the threats that all the penguins face become more immediate. I make a small promise to keep an eye on Turbo. It is only fair, since he keeps an eye on me, stopping at the edge of the sidewalk and checking to make sure I am right there with him. As if to say, *Come, let us go together.*

2
The Point

Much as the penguins have conspired to push it from our minds, a world still exists outside their colony, and we make our first foray back to it at the end of the week. Of civilization's many enticements, what I actually look forward to most is the chance to wear colors other than green or brown. Resplendent in an orange shirt and blue jeans, I fairly strut down the streets of Trelew. No one notices.

We have come to town to pick up two volunteers, but before Ginger fetches them from the airport, she drops El and me off at a huge supermarket called La Anonima (literally, The Nameless). Her single order for us is to shop massively. We are left to shuffle up and down the aisles, wrangling two carts apiece. What we buy doesn't matter so long as we get three or four of them, and a cashier visibly deflates when we approach her checkout with teetering loads of flour, milk, yogurt, granola, tomatoes, *zapallitos* (squat zucchinis), green peppers, apples, oranges, pears, soap, string, napkins, paper towels, hot sauce, mustard, cookies, crackers, chips, cheese, salt, sugar, pepper, spices, dulce de leche, soup, apple juice, orange juice, peach juice, coffee, bread, beef, chicken, olive oil, and other necessaries. Twenty minutes later, after she has rung up a month's worth of provisions, she goggles at the wads of cash we pile before her.

In the parking lot, we meet Briana Abrahms and Emily Wilson, the new volunteers. They are an apt if arbitrary pair: Briana taller and Emily short; Briana more earnest and gentle, Emily animated and forthright. Both have just graduated from college and so seem fresh and shiny. The next morning, Ginger puts them through their paces: the clothes, the gear, the *gancho*, the first hopeless survey of Cañada. El and I watch from afar as the two of them reel around in the bushes, plainly lost, while Ginger observes impassively. We share a sympathetic glance between ourselves, hardened veterans that we are after all of a week.

Now that we are five, Ginger says it is time to add another component to the penguin episteme: a beach survey. Searching for banded birds in the different nest areas is the project's fine detail work; beach surveys address broader, more summary matters. As a task, a beach survey is outwardly simple—walk the beach, count all the penguins, keep an eye out for ones with bands—but this survey, Ginger explains, will be an oiled bird survey. Unlike the regular beach surveys, which we will do at least once a week in the months to come, oiled bird surveys take place only at the beginning, middle, and end of a season. They are made up of three sections, cover-ing the extent of the coastline: North Beach, South Beach, and between them, The Point. Ginger says each section has its merits, and asks which one El and I would like.

"We'll take The Point," I say.

"Fine," Ginger says. "But just so you know, it's the longest section, and you probably won't see too many penguins."

"That's okay," I say. We have both wanted to visit The Point since we got here, in part because it is the colony's namesake, but also because of its remove. The Point holds itself aloof, and this gives it a mysterious allure.

⌐

The next morning, we are all up early. After the Cañada check, everyone moves quietly about the kitchen. We pack extra food, extra water, extra layers. Ginger gives El and me our final, somewhat ominous instructions: be careful on the rocks, don't walk too close to the sea cliffs, and do not, under any circumstances, get within thirty feet of the sea lions, or worse, between them and the water. With that, she sends us on our way.

El and I set off down the tourist trail. The day is bright and clear and cold. We have yet to enjoy much of the desert's promised warmth, but the penguins are at least becoming more boisterous. Almost all the males have returned, and the females are beginning to join them. Their arrival marks the start of the breeding season in earnest. The colony is louder now, the males' brays more fervent, but the fervor is not uniform. Some of the older, more established pairs have a relaxed, almost phlegmatic air. They may have mated together for ten years or more and used the same bush or burrow for almost as long. Some of their nests have become historical repositories for the baubles of Dee's science, with old metal nest markers and pieces of tarp strewn among the twigs and grass.

A male and female preen one another outside their burrow. (Photo by Briana Abrahms)

For those males that are either unattached or have only had a mate for a year or two, the colony becomes a large, open-air meat market. They stridently hawk their wares (my exquisite nest! my gorgeous body!) for the females' perusal. The sex ratio at Punta Tombo is heavily skewed toward males—there might be fifty thousand more of them—and females have the privilege of choice, although to call it a privilege might not be wholly correct. Female penguins are constantly harassed. They can barely waddle ten feet without having a male or two sidle up to them, nibbling at their neck, patting them with their flippers, soliciting, soliciting, soliciting.

El and I walk through this fray, duck under a thin wire fence at the end of the tourist trail, and scuttle behind the *jume* bushes. (This early in the day and the season, no tourists are here yet, but they will come soon, and it will be easier for everyone if they do not see us wandering off-trail.) The survey will trace the shoreline from here to the tip of The Point and back along the other side. El and I are to pace in one-hundred-meter segments while, with a little hand-held counter, we *click* for each penguin we see. We start and soon fall into an easy rhythm, pacing out distance, *click*ing penguins, pacing distance, *click*ing penguins. As Ginger promised, we do not see many. In the first hundred meters we count seven, and in the next, thirteen. Four in the next, only two in the one after that, and then none. None, none, none, and none.

We hop over the rocks, leaving the hordes of penguins behind, trading their urgent noise for the sough of the sea, the rip of the wind. The shore, so craggy and sharp at first, flattens into a smooth table that is easy to walk across. In the absence of penguins, I look to the sky for birds, something I have not had the mind to do since we have been here. Kelp gulls drift over us in loose affiliations, while tight flocks of oystercatchers race up and down beneath them, whistling their high, thin *wheeeeeeeps*. A beige caracara called a *chimango* perches in a bush, a stark silhouette. Past that, I see a single male penguin standing on the rocks, gazing at the ocean. He is something of a surprise out here all by himself, and I gawk at him for a moment before El gives me a little nudge and I remember why we're here. *Click.*

Penguins are creatures of aggregation. Even as they bicker and squabble, they are drawn to each other. In the densest parts of the colony, their burrows might be only a few feet apart, just outside what Ginger calls "snapping distance." A lone penguin thus invites a different scrutiny. El and I watch him, this bird who has apparently chosen to exile himself from the colony for a time, if "exile" is the word I want. Knowing what I know so far about penguins and their fidelity to a place, it might be truer to say this penguin is less an exile than he remains sensitive to the dimming past. In the early 1980s, when Dee started working at Punta Tombo, the density of penguin nests on The Point was much higher, but the years passed, and now only a few are left here and there.

The penguin spies us and straightens, bobs his head. Penguins look so ageless that this bird might be more than thirty years old, old enough

A lone penguin in the pampas grass on the way to The Point.

to remember when this land was more thickly occupied. He might even have hatched out here. Maybe he likes to slip away from the madding crowds for a few hours to get a little peace and quiet, to reclaim this open country for himself. Or not. I am probably making too much of the bird. As Ginger has often told us, penguins have brains about the size of a peanut. (She likes to demonstrate this with her fingers.)

The penguin eyes us as we get closer. He crouches, ready to hurl himself into the sea, but when we do not approach him, he relaxes. When we pass, he turns his back to us and resumes his gazing, as he might have been doing for ten minutes, or for much longer.

∽

We walk on, the lone penguin becoming smaller behind us, now a speck, now indistinguishable from the rocks, but for me his mystical stature grows. To see a Magellanic penguin in the deserts of Patagonia is to see penguins as they were originally conceived, in a way, for this was the species that introduced the idea of *penguin* to the wider world. (Or, more accurately, to the European world, bearing in mind George Gaylord Simpson's observation that we don't think a thing exists until a European has seen it, and "even then we question the matter until the 'discoverer' has returned home and written a book.")

The introduction was brief, and overshadowed by events of larger significance. It happened in or around 1525, when a Parisian bookseller published the journals of a young Venetian named Antonio Pigafetta. Pigafetta was one of eighteen men to survive Ferdinand Magellan's attempt to circumnavigate the world, and one of only four to complete the entire trip. His presence on Magellan's crew had been something of a happy accident. In 1519, while in the city of Seville with the papal ambassador, Pigafetta was in attendance at the Mass where a representative of King Charles I bestowed the Spanish royal colors upon Magellan, granting the king's blessing for the Captain-General's upcoming journey. Pigafetta was entranced. Flush with "a craving for experience and glory," as he would later write, he offered Magellan what services he could. He bore letters of papal recommendation, and those, along with his willingness to take a low salary, secured him a place on the crew. His official role was listed as "supernumerary," which seems to be a polite way of saying he was along as a tourist. Mindful of this, Magellan gave Pigafetta a special assignment: he would keep a record of all that happened on the voyage. It was not to be a dry account in the manner of a ship's log, but a piece of travel literature, personal and lively.

Pigafetta accepted the charge. His eye and pen would be both wide-ranging and wryly distanced. "If Magellan was the expedition's hero, its Don Quixote, a knight wandering the world in a foolish, vain, yet magnificent quest," one historian wrote, "Pigafetta can be considered its antihero, its Sancho Panza, steadfastly loyal to his master while casting a skeptical, mordant eye on the proceedings."

One day in January of 1520, Magellan and his crews were sailing down the South American coast. "Then, following the same course toward the Antarctic Pole, coasting along the land, we discovered two islands full of geese and goslings and sea wolves," Pigafetta wrote in his journal. "The great number of these goslings there were cannot be estimated, for we loaded all the ships with them in an hour. And these goslings are black and have feathers over their whole body of the same size and fashion, and they do not fly, and they live on fish. And they were so fat that we did not pluck them but skinned them, and they have a beak like a crow's."

In his entire account of the expedition, this is the only passage in which Pigafetta mentions the animal that would come to be known as the Magellanic penguin. Of special consternation is that, despite his otherwise meticulous attention to detail, he did not note where the ships were when the geese were first seen. He wrote only that they were somewhere south of the Rio de la Plata, which is near what is now the city of Buenos Aires.

A mild dispute arose over where exactly they were. Some scholars argued Magellan was in the strait near Tierra del Fuego that now bears his name. (The Magellanic penguin is actually named for the Straits of Magellan, rather than for Magellan himself.) Robert Cushman Murphy, a prominent American ornithologist in the early twentieth century, thought the crew was farther north, in the Golfo San Matias near the Peninsula Valdés, roughly two-thirds of the way down the Argentine coast. George Gaylord Simpson, who as a paleontologist had worked extensively in Patagonia, disagreed. When he compared Pigafetta's notes with those of Francisco Albo, Magellan's pilot and keeper of the official log, he determined the ships were just south of the Peninsula Valdés, closer to 44°S latitude. It is a place with a few small bays and a small spit that juts into the sea. "This is, indeed, almost exactly the latitude of Punta Tombo near which there is a large penguin rookery evidently of considerable antiquity," Simpson wrote. "It is a good guess, or rather more than a guess, that Pigafetta's notice that gave the news of penguins to the European world referred to 27 January 1520 and to this part of the coast."

El and I continue along this part of the coast, with nary a penguin in sight. To our left is the ocean, to our right equal parts pampas grass and sky. The margin between them is flat and unbroken. Clouds rush over us. The scene is empty, lyrically so. Then the peninsula starts to narrow, and a slim wedge of sea appears to the south. It enlarges as if compressing the land, and before we realize it, we have arrived at The Point. There is nothing in the way of shelter here and the wind is fierce. Rocks jut up through the sand like jagged teeth. The northern side of The Point slopes gently to the Atlantic, but the southern climbs until it ends abruptly in a cliff some twenty or thirty feet tall. The sea leaps and crashes against it. Above us, hundreds of South American terns wheel in a shrill, whimsical storm.

Now that we are here, we venture forward more cautiously, taking care not to disturb anything—the animals, the mystique. On the very tip of The Point is a mixed colony of imperial and rock cormorants. We count them—this being a survey, counting is reflexive—and find there are roughly one hundred and twenty imperial cormorants and fifty rock cormorants. Compared to the penguins, it feels pitiful to call so few birds a colony; or maybe it is generous to grant them such status.

The small mixed colony of a couple hundred imperial cormorants (*Phalacrocorax atriceps*) and rock cormorants (*P. magellanicus*) on the farthest tip of Punta Tombo. White bellies distinguish imperial cormorants, while rock cormorants have entirely black plumage.

(Decades ago, this cormorant colony numbered in the thousands.) They are surrounded by attendant predators, keeping watch for their eggs: kelp gulls, dolphin gulls, two or three pairs of hulking Antarctic skuas, and a few strange white birds called snowy sheathbills. The air between kelp gull and cormorant is languid but tense, as it tends to be when a consumer and the consumed must tolerate each other in uneasy détente.

El and I stop and stretch, and I ask her what she thinks of all this: *this*, the penguins; *this,* Punta Tombo. The beach survey is the first chance we've had to reflect when we are not spent at the end of the day; our chats in the trailer before we fall asleep are brief and gnomic. Given the opportunity to expand, El is noncommittal. Working with penguins is fascinating and their colony is awesome, we agree, but awesome in every sense of the word. Our feet ache, my knees are grinding stones. It is unnerving that we have only been here a week and already are so tired. More than tired: fatigued. Six months of this labor positively looms.

"At least it's pretty," I say. El makes a face. She misses the mountains and trees, the fresher green of home. This desert does not speak to her, even though it is so loud.

"It'll get better," I say, and this uncertain promise brings our discussion to a close. How could I be so sure it will get better? What information am I privy to that El is not?

A short distance from us is a large rock, maybe eight or ten feet tall. It is the highest thing here and looks warm in the sun. "Let's climb it," El says. "We can eat our lunch there."

We clamber up the rock and swing our legs over the edge. Lying at its base not five feet from us is a young bull sea lion we couldn't see from the other side. It is well within the thirty-foot radius Ginger had stressed we should maintain between us and any living sea lion, and I remember her other name for the area inside such a circle: the Zone of Death.

The sea lion is asleep, but it snorts awake when our shadows fall across its face. It stares up at us. We stare down at it. If we are looking at something we have never been this close to before, then so is he. He bellows at us, and whether out of fear or anger his voice seems to rise from deep in the earth, amplified by his great chest. His guttural roar passes right through me, taking some part of my spirit with it. Without thinking, El and I leap from the rock and run. The sea lion galumphs away in the opposite direction. Fear, then. It is afraid of us, too.

Shaken, we make our way over to the edge of the cliff on The Point's southern side, which overlooks a large punchbowl. Our mulling of Punta Tombo is over; we have more pressing matters to discuss. What if the sea lion comes back? Will it attack us? Do sea lions attack people? I don't think so, but anything seems possible here. I can't help but look over my shoulder every few seconds, expecting to see the enraged young bull bearing down on us to slake his bloodlust. If we threw ourselves into the sea, would we survive? Probably not. The waves smash the rocks with a rollicking force, and my imagination is playing through a series of panicked scenarios when, amidst the tumult below, I see a single penguin.

It swims calmly, with easy skill. A wave crashes over it and it shakes its head, spraying a halo of drops. Another wave bears down on it, but it ducks into this one to meet it, and puckishly pops out once the wave has passed. It idly paddles about, buffeted by the surging ocean, batting aside the sea's power. When it has seen all it wants to see, it dives with a few quick, sure strokes. It streaks away just beneath the surface, a swift shadow in the roiled sea. I follow it as far as I can until its form melts into the violent green glow of the water, and it disappears. Mesmerized, elated, I reach for my clicker before I forget. *Click.*

⌒

After Pigafetta, Magellanic penguins would become a common feature in annals from the Age of Discovery. They were a reliable food source, and Penguin Islands were marked on maps to indicate colonies suitable for harvest. (Large tame flightless birds have never fared well with the explorers and colonists who found them.) Sir Francis Drake, in 1578, wrote of the "great store" of penguins on Santa Magdalena Islet; his men killed at least three thousand there in a single day. In 1586, another English sailor and privateer, Sir Thomas Cavendish, landed at Puerto Deseado, about four hundred miles south of Punta Tombo. He and his crew killed several thousand penguins for "victuals." They were back again in October of 1591, readying for the return sail to England. After two months of butchering and salting, they loaded fourteen thousand penguins into the ships' holds. (Cavendish had estimated that four men would need five penguins per day for the six-month journey, and he had to provision four ships.) A few years later, the crews of the Dutch sailor Oliver van Noort claimed to have killed fifty thousand penguins, while also collecting "eggs innumerable."

Magellanic penguins could provide amusement in addition to sustenance. In 1594, Sir Richard Hawkins, an English explorer and pirate, visited the same colony Drake had visited. There, he set about filling "some doozen or sixteene Hogsheads [large casks]" with penguins, which he deemed "reasonable meat rosted." To procure their vittles, Hawkins's men armed themselves with cudgels, herded the penguins into the center of a great circle, and bashed them on the head. "The hunting of them (as wee may well terme it) was a great recreation to my company and worth the sight," Hawkins wrote. Especially fun was when a few desperate birds tried to escape. Then "was the sport," as penguin and human alike raced over the burrow-ridden ground. The penguins' nests were hidden deep in the tall grasses, and the men crashed into them, tumbling over one another while the penguins scrambled up the rocks to get away. From then on, the penguins gave humans a wide berth: "And after the first slaughter, in seeing us on the shore," Hawkins wrote, "they shunned us, and procured to recover the Sea . . ."

Three hundred years after Magellan sailed past, Charles Darwin himself would sail along the eastern coast of South America as the naturalist aboard the *HMS Beagle*. He had first seen Magellanic penguins in August 1832, in the Rio de la Plata estuary, near Montevideo, Uruguay. They surrounded the *Beagle* one night and seemed to glow, he would write in a letter to a friend, "darting through the water, leaving long trails of phosphorescence in their wake." Later, in December 1832 and again in 1834, he saw them when he visited the Falkland Islands, or Islas Malvinas. Struck by the penguins' character and pluck, he tried a small experiment. "Another day, having placed myself between a penguin (*Aptenodytes demersa*) and the water, I was much amused by watching its habits," he wrote. "It was a brave bird; and till reaching the sea, it regularly fought and drove me backwards. Nothing less than heavy blows would have stopped him; every inch he gained he firmly kept, standing before me erect and determined."

As it had with other explorers, the Magellanic penguin's ecstatic display made an impression on Darwin, but he also was able to see beyond the dogma of penguins as loud and comic animals, sensitive as he tended to be to shades of phenomena most others missed. "The bird is commonly called the jackass penguin," he wrote, "from its habit, while on shore, of throwing its head backwards, and making a loud strange noise, very like the braying of an ass; but while at sea, and undisturbed, its note is very deep and solemn, and is often heard in the night-time."

After El and I finish lunch, and once we have made sure the sea lion is nowhere nearby, we start back. The rest of the survey will follow The Point's southern side. As before, penguins are largely absent, but the walk is no longer so quiet and peaceful. A large colony of kelp gulls abuts The Point. Hundreds of them whirl around us as we get closer, calling and carrying on. I note with some dismay the gull early warning relay system: those we have passed land after a spell, only for those in front of us to take off, so that we are constantly accompanied, constantly harangued. To look for penguins here seems absurd. None could be so dumb as to live in the middle of so many kelp gulls.

Once we have escaped the gulls' sphere of influence, the scene stills. The mainland may be closer now, but I am looking at the ground, inspecting each speck. Bones are everywhere. They draw my eye. Some are so large they must be the bleached vertebrae of whales. We see the decaying carcass of a sea lion, its angry red skin stretched taut across its face, pulling its lips back into a desiccated snarl. A bullet hole puckers its back. (Fishermen sometimes shoot sea lions to eliminate the supposed competition.)

The vast majority of carcasses are birds, and most of those, penguins. Ginger had said we should measure and weigh the ones dead for a month or less, but we are not yet experienced enough to tell when this or that bird died, so we leave the bodies alone. Sprawled across the rocks and sand, a few of the fresher ones look like they could be napping. Others have been here longer and are more conventionally disgusting. The oldest are just half-buried bone, held together by threads of sinew. Stubborn clumps of feathers cling to them. I pause over one skeleton. It is stripped of all gristle, and the bones gleam white. Their arrangement suggests an artistry more deliberate than mere skeleture. A small pink flower peeks out of a jumble of ribs; the body is folded about the bloom, sheltering it. The bones blend with the rocks and white shells, and the wind blows through them, but gently. It stirs the flower, stirs the sand, and in time the sum of these stirrings will abrade the little diorama, until this penguin is dust once more.

⌒

Darwin left Argentina in 1834, and the *Beagle* sailed on around the tip of South America and north to the Galápagos. The time he would spend among those islands, and the book that would come more than twenty years later, are what he is most remembered for now, but the Galápagos were not what he himself thought of near the end of his life. "In calling up images of the past," he would write, "I find the plains of Patagonia frequently cross before my eyes: yet these plains are pronounced by all most wretched and useless. They are characterized only by negative possessions; without habitations, without water, without trees, without mountains, they support merely a few dwarf plants. Why then, and the case is not peculiar to myself, have these arid wastes taken so firm possession of the memory?"

⌒

El and I pace now in silence. My gaze wanders out over the pampas grass, across the wan sky. Even after only a few days, I feel the pull of this land. It possesses you, claims you, makes claims of you. We had been warned this could happen. Days before, while El and I were in the parking lot at La Anonima waiting for Ginger and Briana and Emily, two young boys had come up to us and demanded to know where we were from. (We were obviously tourists of some kind.)

In Spanish, El told them we were from the United States.

"Is it pretty there?" one of the boys asked.

"It depends on what part," El said.

"Is it prettier than here?" the boy pressed, waving their hands to indicate *here*—Trelew, Patagonia, El supposed. She thought of what we had left behind: the Pacific Northwest, the Cascade mountains, with their wet dark forests and cold clear lakes. *Yes, yes,* she wanted to say, *emphatically yes, where we came from it is so beautiful it takes your breath away.* But she understood that was a different kind of beauty. Here she had other obligations. She could tell these boys may have been bored with Trelew, a dusty outpost in a vacant land, but they were also proud of their home. She shook her head and smiled. "No," she said. "No, it is very pretty here. Patagonia is so beautiful."

The boys grinned, satisfied, and ran off.

We reach the end of the survey at a large rock formation. It vaguely resembles a frog, or so Ginger says; in any case, Dee calls it Frog Rock. This is the beginning of the South Beach section, which Ginger did on her own. One penguin is lying on the sand. When it sees us, it jumps to its feet and bolts for the sea, using its flippers as forelegs and scrambling on its belly in a fashion called tobogganing. It flails into the surf and pops up a few yards from shore, sculling in the water, never taking its eyes off us. This is a penguin that knows humans and our appetites only too well.

I reach for the clicker—*click*—and look at the results. After covering almost five miles, we have seen seventy-three penguins.

"Well, that was fun," I say.

"Yes," El says.

Now that we have finished, we head toward the house along the colony's outskirts, taking care to stay low so the few tourists who are here today will not see us. When we come to a draw, we follow it down toward the main trail. It drops us into one of the study areas, Doughnut, I think. Penguins again surround us with their honking and braying, but after that little reprieve it is nice to see and hear them again.

We walk up the trail, idly scanning nests until El pauses on one in a small *ojo de vibora* bush that is somehow odd. We can't say what makes this pair of penguins different from all the hundreds of other pairs in hundreds of other nests we have walked past. It might be the way the female is standing. She crouches and seems furtive, as if she is hiding something.

We approach the pair. Both the male and female furiously head-wag at us. El kneels and gives the female a light nudge with her *gancho*. The female steps back and glares, and El sees what she has been trying to conceal. She looks over at me and smiles. "It's an egg," she says.

3

Hurricane Dee

Not long after the oiled bird survey, I step outside the house one evening and hear two penguins fighting nearby. It is early October, and fights have become a more common feature of the Punta Tombo soundscape now, the improvised rage of battle often bursting through the formulaic howls of the ecstatic displays. Penguins will fight over nests, over mates. The battles are straightforward, if inelegant: the birds simply beat each other with their flippers and rip with their bills until one, usually the smaller of the two, has had enough and runs away. The worst fights happen when two males of equal size feel they have an equal claim to whatever it is they are fighting over. Their bouts might last half an hour or more. Afterward, the two combatants will sometimes lie next to one another, panting for breath, their heads and chests and flippers soaked with blood. Death seems imminent, but really all they need is a trip to the sea to rinse off. When I next see them, they look little the worse for wear, save for some cuts and scabs and a certain puffiness about the face.

The penguins may suffer no lasting harm, but that doesn't make their violence any easier for me to stomach. I move to avoid this fight when I see it is coming from a bush, and then I realize it is Turbo's bush, and then I realize one of the fighters is none other than Turbo himself.

"Turbo is fighting!" I cry, and everyone rushes from the house. None of us can make out anything through the clouds of dust the penguins kick up. All I hear is their shrieking, and the solid *whumpwhumpwhumpwhump* of one walloping the other. I shuffle in place, flutter, fret. Turbo became who he is because of a fight. What if he gets smacked on the head again and forgets himself? What if he stops visiting us?

Suddenly I hear some triumphant braying, and a penguin rockets from the bush. Turbo is right behind, snapping at the intruder's tail. He skids to a halt, huffing and puffing, and then gathers himself. He is in a bad state. He favors his left foot, his breast and flippers are smeared red, his face is deeply gashed, and blood is pooling under his right eye. But he has won. He limps past me to the front of his bush, ignoring my frantic ministrations. He rears back, throws his red flippers wide, and brays: *Hu-hu-hu-hu-hooAAAAAAAAAAAH! HooAAAAAAAAAAAH! HooAAAAAAAAAAAH!* In the full throes of his penguin-ness, he brays loudly, defiantly, splendidly. He brays again and again and again. When I go to bed an hour later, he is still braying.

⤛

The Battle of Turbo, as I come to think of it, underscores the curious energy about the colony these days. We can all feel it. For Turbo, for the penguins, it is the all-consuming imperative to breed. We are less sure what it means for us, but Ginger has started to drop hints: we should

Two male penguins fight over a burrow while a female looks on.

build up our strength, get plenty to eat, rest up. "You need to be ready," she says, "for when Hurricane Dee makes landfall."

We all titter. "Hurricane Dee?" El says. "Did she come up with that name herself?"

"No," Ginger says. "It was one of her students a couple of years ago." She grins. "I'm not even sure Dee knows people call her that."

Then we are all off to the airport again, me in my wildly colored clothes. "Don't get used to this," Ginger says as we watch the passengers deplane, the giant plaster penguin glowering down at them. "When Dee is here, we won't come to town for at least a month."

Dee bursts from the baggage claim. An inflatable neck pillow dangles from her shoulder, and she lugs a duffel bag like a piece of downed timber. The bag is filled with equipment she brought from Seattle, but also, she promises, dried peaches and mango that she dehydrated for us. She stows her luggage in Opus's bed and wedges herself into the backseat between Briana and Emily. On the drive back to the colony, she tells us about life in the United States, but the financial crisis and upcoming presidential election are of little interest to her. In the past, she spent months at Punta Tombo, before the duties of being a penguin maven limited her to only two or three visits each year. Of course this is hardly enough time, and she has to make the most of it. She peppers us with questions.

"How is everyone doing?"

"Fine," Ginger says. "We're a little tired."

"El? How are you finding this science thing?"

"It's interesting," El says.

"The house? The trailers? How are they holding up?"

"They're fine," Ginger says. "No major leaks, and no problems with the solar system."

"And the penguins? How were they?"

"They seem to be doing okay," Ginger says.

"Good. Is Packy here?" (Packy is one of the first penguins Dee ever banded.)

"Yep. He came in pretty early."

"Great. Does he have his old nest?"

"He does. He even has a mate."

"He's a real trooper. He will never die. Is Turbo back?"

"Yes, and he still comes to the house."

I pipe up: "He got into a huge fight a couple of days ago!"

"Ha! Does that mean he got a female to join him this year?"

Ginger laughs. "Of course not!"

"But he's otherwise fine?"

"He's his usual dapper self."

"Great, great." Dee settles in and is quiet for a moment, gazing happily out the window. Bushes blur past. Opus flings stones. Dee leans forward and rests her elbows on the front seat. "Think you can drive any faster, Ginj?"

Dee may now be linked with the Magellanic penguin in Argentina, but the linkage was hardly foreordained. She grew up far from the ocean, in Ann Arbor, Michigan, and as a child was interested in moths and butterflies. These she pursued around her backyard with characteristic vigor, stamping the wings of the ones she caught with a little ink so she could follow various individuals. It was a fun hobby, but that was all. Her mother was a high school teacher, and her father owned a travel company; both expected she would follow in one or the other of their footsteps.

Dee was attracted more to the path of her grandfather, who at the time was president of Central Michigan University. She matriculated there as an undergraduate student a year after he retired and majored in biology. When she graduated, still intent on becoming a college president, she asked her mentor whether it mattered what she got her PhD in. (She assumed she would need one to be an academic administrator.) He said it didn't, and so she decided to continue in biology. If she was going to treat a PhD as a means to an end, she might as well get one in something she enjoyed.

In 1970, Dee was accepted into the lab of Paul Colinvaux, a biologist at the Ohio State University. Colinvaux was best known for his research in the Arctic and the Amazon, but he had done projects in the Galápagos, and Dee had always wanted to go and see the life that had given shape to Darwin's theory of natural selection. When Colinvaux asked her what the focus of her dissertation would be, she plucked the Galápagos penguin more or less out of the air. (In part she thought it odd, given the popular perception, that a penguin would live on the equator.) "Wonderful," Colinvaux rumbled. He was British, and a big man; to Dee, he sounded like Winston Churchill. Fortunately for her, he was also one of the few academic biologists willing to send a female graduate student out into the field by herself.

The following June, Dee was on a plane to the Galápagos, six hundred miles off the coast of Ecuador. Once at the Charles Darwin Station on Santa Cruz Island, she asked around and learned the penguins were actually elsewhere, on Fernandina Island. She found some geology graduate students who were going to Fernandina to study a volcano, and said she would pay her way if they took her along. They agreed, and after a two-day boat ride they deposited her, alone, on the tip of Fernandina, called Punta Espinosa. She had a tent, a tape measure, a notebook, some pencils, enough food and water to last two weeks, and a growing appreciation for what exactly she had taken on. She was twenty-two. She had never even camped before.

She set up her tent and went for a walk to explore the island. The ground was hot and dry, the terrain otherworldly. All the trees were bare. Although thousands of marine iguanas and dozens of sea lions bobbed about the shore, she did not see a single penguin. Since the Galápagos penguin is a full-throated character like the rest of its jackass cousins, she was worried, but at night, back at her tent, she heard a single male braying off in the distance: *Hu-hu-hu-hooAAAAAAAH!* She thought it a lonely, plaintive sound.

Despite such ambiguous omens, Dee went to work. She set off across the baking black lava to search for penguins, and found them stashed in tubes, crevices, and caves, where they were sheltered from the equatorial heat. She wrangled more than four hundred adults and juveniles and measured them: their flippers, their feet, the lengths and depths of their bills. When the females laid their eggs—like the Magellanic, they lay two—she measured the clutches. When the chicks hatched, she measured them, and returned every few days to track their growth until they either left the nest or died. To census the population, she took a boat out around the island, tallying penguin bills, or "noses," and counted nearly two thousand. From this, and her daily counts around Punta Espinosa, she estimated the total population on Fernandina and Isabela islands to be between six and fifteen thousand birds, making the Galápagos penguin one of the rarest species of all.

⌒

Over the next two years, Dee would spend three hundred and fifty days on the island. The animals were so tame she sometimes felt like she was living in a petting zoo. Sea lions slept against her tent; their snoring would keep her up at night until she nudged them and they quieted down. Then there was her work. So little was known about the penguins' lives that almost everything she learned was new information. It was natural history at its most exhaustive and intimate. With it as a lens, she was able to see links between the sea, the climate, the birds—connections that would inform the rest of her career.

The marine life of the Galápagos is governed by three major ocean currents: the North Equatorial Current, which brings warm, tropical waters from Central America in the northeast; the South Equatorial Current, which carries cool, subtropical waters up from Antarctica via the Peruvian Current and its attendant upwelling zone along the South American coast; and the Equatorial Undercurrent, also called the Cromwell Current, a jet of cool water from the west that runs about three hundred feet below the sea's surface, tracing the equator in the Pacific Basin, a distance of some thirty-five hundred miles. The archipelago sits at the confluence of these currents, and its seasons are marked more by changes in ocean-driven precipitation and cloud cover than temperature, which on the equator is fairly constant. From roughly June until November, when the southeast trade winds are strongest, the South Equatorial and Cromwell Currents have the most influence. The ocean is cooler. Little rain falls. On land, this is the dry season, known locally as *garúa*, for the mists that wreathe the highlands. When the trade winds shift in December and come from the northeast, the warmer North Equatorial Current dominates. Both ocean and land heat up, and the rains come. This is the wet season, which lasts until May.

The land may be lushest during the wet season, but Dee saw that marine life thrived in the drier months, when the cool, nutrient-rich waters of the Cromwell Current spread throughout the islands. They brought plankton and great schools of fish. The islands' seabirds—boobies, frigate birds, pelicans, shearwaters, noddies, penguins—descended on the schools in mad feeding frenzies, some plunging headlong into the sea from as high as one hundred feet in the air. During one frenzy, the birds dove with such abandon that they made the water boil for five hours. Dee watched from shore, astonished. The sea looked black, so dense were the fish. She could scoop them from the water with her hands.

The climate pattern of wet and dry seasons generally held, but there were exceptions. A season might be shortened, or lengthened, or sometimes skipped altogether. The ocean varied as well and was subject to irregular shifts. The most significant of these shifts, called El Niño or The Child—in South America its effects are strongest around Christmas—happened when the southeast trade winds weakened at times they should have been strong. Wind drives water, so the cooler currents were no longer blown to the Galápagos. The sea around the islands calmed and warmed. Without the nutrients that cooler water brings to support them, the plankton disappeared, and with the plankton went the fish. The seabirds thus faced dire food shortages. Many starved.

Dee wanted to know how the Galápagos penguin had adapted to such a fickle environment. She learned that, unlike birds whose lives are more seasonally fixed and predictable, these penguins were flexible. Rather than time, they responded to the amount of available food, breeding only when there was enough, taking what the ocean gave them. They nested whenever they could, at any time of year. More than half of them would breed twice in a single year, and a few even bred three times. (Most seabirds breed once a year, and some, such as albatrosses, only breed every other year.) They could also molt at any time, but always before they bred instead of after, like all other penguins. Dee thought this showed how sensitive they were to the chance their food might vanish: they took care of themselves first, and then tried to raise offspring with whatever resources were left. Penguins, she was starting to see, were more than just penguins. They were indicator species, sentinels of the oceans. Their behavior, their lives and deaths, traced global phenomena.

As much as she loved the Galápagos penguin, though, studying such a rare bird started to get dreary. The archipelago was climatically dull, either hot and wet, or slightly less hot and dry. She wanted a different kind of abundance, wanted the subtler hues of other seasons. She had finished her PhD by then and was a young professor at the University of Washington in Seattle. In 1976, she went to the Barren Islands in Alaska to study fork-tailed storm petrels. Fork-tailed storm petrels are lovely birds, small and sprightly and quick as they flit over the waves. To work with them offered a reprieve from the desert heat, and, if Dee was honest, it was nice to spend some time with seabirds that fly.

⌒

The egg El found on our way back from The Point was the first seen this season, and Dee's arrival has coincided with their appearance wholesale. Female Magellanic penguins lay two eggs, four days apart. The first is usually longer and more conical, and the second shorter and squatter, although both have the same volume. Each is about half again as large as a chicken egg and white or, if you happen upon one within minutes of its laying, the lightest blue. Soon, they will be coated with dirt, guano, blood, other mysterious substances. Nothing stays clean for long at Punta Tombo.

The male penguin has keenly awaited the eggs. He has spent the last month or so at his nest, unable to leave for fear of losing his home or his mate to one of the colony's many eager lotharios. During this time, he has lost several pounds. When he backs tail-first to the entrance of his nest to defecate, he squirts out sticky dollops of green guano: the bilious waste of an empty stomach. When the first egg comes, then, off he goes. He will swim north and might travel more than four hundred miles to look for food. He won't return for up to three weeks. Even males without mates feel the pull of eggs. We sometimes find them in their nests sitting on white rocks or other eggish objects, frantic slaves to their thwarted reproductive hormones.

With many thousands of males now gone, and many of the rest preoccupied with rocks, the colony becomes so quiet it seems deserted. Wedged in the back of a burrow, it is now a hushed female that head-wags at El and me when we come to see if she has an egg for us to measure. Like her mate, we have also been waiting for it. Dee wants to know within forty-eight hours when known-age penguins (or their mates) lay eggs, so we have been checking this nest every other day since we first saw this female.

I drop to my knees and peek into the burrow. In one hand I hold my *gancho*, in the other an egg cup. Like all the best field equipment, the egg cup is a marriage of ingenuity and domestic salvage: we have taken a large tin can, mummified it in duct tape, and lashed it to a yardstick. For all its apparent usefulness, I feel ridiculous lugging it around, especially when I have to explain to bemused tourists in my caveman Spanish, with supplemental gestures, that I am off to "touch the eggs with my cup."

"You really should try to learn more Spanish," El laughs after one such encounter. "It would be good for you."

"Yeah, yeah," I mumble. I do not have her facility with language.

To retrieve the egg, I thrust the egg cup into the burrow. In that confined space, the cup compels the female to stand. She had sheltered her egg under a strip of bare pink skin called a brood patch. (She has plucked her own feathers to expose the skin, which is where some of the blood speckles come from. That, and the fleas.) I maneuver the egg into the cup with my *gancho* and draw it out, taking care to keep the egg safe from the female's defensive jabs. I measure the egg's length and width and hold it in my hand to feel its relative temperature. It is warm, almost hot; an incubated egg might reach nearly 100°F. Finally, I scribble *#1*s all over it in black Sharpie so we will know which egg it is on future visits.

I next haul out the female so we can weigh her. Her weight will tell us whether she has enough fat to last out her mate's absence. A fasting penguin loses about two ounces per day, and in good years, females will weigh more than nine pounds between the first and second egg. This female weighs a little more than seven pounds. The timing for this pair is going to be close. Dee has found that once a female's weight drops below 6.1 pounds, she will almost certainly desert her clutch and go feed herself.

We release the female back into her nest. She scuttles in, flippers trembling. I now have to return the egg, a task complicated by a quirk of penguin cognition. Once an egg is no longer in the nest cup—once it is merely outside the range of the penguin's bill—the penguin will no longer recognize its own egg as an egg. Instead, the egg becomes like any other Not Me in the nest: an invader. This transmogrification from Fruit of My Loins to Mortal Foe is both abrupt and absolute. As I scoot the egg cup back, the egg rattling inside, the female hammers away, grabs the cup with her bill, bites it, thrashes it. Were the egg to bounce free, she would peck it without a second's hesitation, but once I have returned it, she is quick to forgive her egg its absence. She clucks at it and nuzzles it back under the warmth of her brood patch. Resuming her incubative pose, she regards El and me with a mild look, seeming to wonder why we are there.

⌒

After a few days of this, I tell Ginger I am comfortable enough handling penguins to try doing egg checks by myself. Maybe I am feeling cocky, I don't know, but weighing a penguin alone turns out to be harder than I anticipated. The females may be smaller, but they struggle just as vigorously as the males do when I loop the strap around them. Ginger's assurances ring in my ears—"You are stronger, faster, and smarter than any penguin!"—but I am not so sure. One female nips my finger when I grip her neck too low. As my blood freely drips into the dirt, I marvel at the casual sharpness of her bill. Another manages to twist around in my hand and beat me with her flippers, drumming my wrist. I gasp in pain and drop her. She scrambles to her feet and leaps at my face, sending me sprawling. I decide to weigh her another day.

Not all injuries are physical. The worst rarely are. Once, when I try to draw a female out of her nest on a ledge in the Doughnut area, she drags her bill and catches her egg. I watch, aghast, as she pulls it to the ledge, and then it is over, bouncing down the short rock wall. It falls only a foot or so, and for one desperately happy second I think we have all lucked out. But then I see a small crack at the egg's apex. Its integrity is compromised; the chick developing inside will never hatch, turned instead into black fetid broth by an onrush of bacteria. My face curdles. Stupid penguin, dragging her stupid bill over the stupid ground and cracking her stupid egg. Because this is obviously her fault.

Back at the house, everyone murmurs sympathetically when I relay what happened. Someone suggests putting duct tape over the crack, but there are limits even to what the almighty duct tape can fix. Dee tells me I shouldn't be too hard on myself. Is a penguin that has never met us better off? Perhaps in the immediate, but you have to remember we're here to help this population of penguins more than any one penguin. That is the level at which ecology operates.

What she says is true, of course, but that night, staring at the ceiling while the penguin under the trailer wails away, I wonder if it is possible to study something without disturbing it, to find without harming the found.

꜁

Dee worked with storm petrels for several years before events in Argentina brought her back to penguins. In 1981, a Japanese business called the Hinode Penguin Company had approached the government in Chubut. At the time, the colony at Punta Tombo was thought to have at least one million penguins. One million penguins is way too many penguins, the Hinode representatives argued. They proposed to harvest more than forty-five thousand penguins each year, using the meat for animal meal and the skins for golf gloves and wallets. It would be, they said, a rational culling.

Even though Punta Tombo had been a nature reserve since 1979, the proposal still intrigued some Argentine government officials. Unemployment in the province was high, and the cull would provide sixteen to twenty jobs for about three to four months a year, at least until the colony ran out of penguins. The public was less receptive. When news of the proposal leaked, people responded with outrage. Hundreds wrote angry letters to legislators. Some even marched on the governor's house. "They made their views known," Dee says now. "Somewhat surprisingly, the government listened to them."

In the end the proposal was rejected, but that it had been entertained at all alarmed a conservationist named Bill Conway. Conway was the director of the Bronx Zoo in New York, as well as the zoo's field research arm, now called the Wildlife Conservation Society. He had traveled widely in Patagonia since the early 1960s and first visited Punta Tombo in 1964. He and the society had worked for years to establish the reserve, and he had since been casting about for a researcher to study the penguins there. Doubly committed now, he quickly settled on Dee. While he was duly impressed with the long list of awards and honors she had already won (including, I see with some amusement, an award for diplomacy), he was struck more by her toughness, her energy, her enthusiasm.

Dee first went to Punta Tombo in 1982. Standing on the barrier beach, looking out at the raucous multitude of penguins, she tried to see the land through their eyes. What mattered to them? Some nested in bushes, others in burrows they had dug out of the hard, pebbled ground. Which was better? Was it important to be closer to your neighbors? Far from them? Closer to the sea? She started marking dozens of nests with strips of tarp so she could follow their occupants' progress. She drew lines around a few different types of habitats and gave those areas the quirky names with which El and I are becoming so familiar. (Bungalows, I learn, was first known as "Boersma's Bungalows.")

The next year, she took a sabbatical from the University of Washington and spent the entire season at Punta Tombo with a few students.

The campo at Punta Tombo.

Together they banded thousands of penguins, filled notebooks with measurements of bill lengths and depths, flippers, feet, weights, eye colors. They drew crude maps with burrows and bushes strewn across the pages, as if someone had spilled a cup of false eyelashes. "I thought I would come back for only a few years," she told me once, but I find that hard to believe. She is not so obtuse. She had to see how the questions would pile up, branch out in unexpected ways, lead only to more questions.

Sometimes, when I'm walking around the colony, I wonder what it was like for Dee in those first years, how it felt to look at and listen to hundreds of thousands of penguins and realize no one knew much about them other than that they were here. Once I was in her office in Seattle, scanning all the books on her shelves. I took down one thick and much-loved volume, *The Biology of Penguins*, a collection of articles edited by noted biologist Bernard Stonehouse in 1975. Dee had written the chapter on the Galápagos penguin—"Adaptations of Galápagos Penguins for life in two different environments"—but I flipped to the Magellanic penguin chapter, by J. Boswall and D. MacIver. The first sentence was, "The

Magellanic penguin . . . is one of the least-studied of the world's eighteen penguins." The phrase "least-studied" was underlined in pencil, and the line had a slight quiver to it, as if the hand that held the pencil had been trembling in anticipation.

⌣

Out on the campo, I see better where that quiver came from. Dee is in her sixties now, short and a little stooped after all the years of wearing a vest laden with twenty pounds of gear, but she still moves around with prodigious speed. To the standard template of field garb, she has added her own unique subtleties of dress: a pistachio foam visor over her short thatch of white-blonde hair, sunglasses to shield her sharp blue eyes, fingerless wool gloves to protect hands that are the color and texture of parchment. Her personality, already formidable, somehow enlarges. She goes from being Dee to Dee-and-a-Half. Or, as seems obvious now, the transformation is in the other direction, and it is at Punta Tombo that she is wholly realized, wholly herself: Dee-in-Full.

She radiates affection for the penguins. She finds them endlessly amusing. One of her first days, we happen upon a male and female about to copulate. As an act, copulation is both elaborate and frequent. It goes like so: A male and female might bill-duel, and then the male parades around the female in a circle dance, lightly drumming her body with his flippers. She lowers herself to her belly and lifts her tail. He climbs onto her back, balancing unsteadily while kneading her rump with his feet, and brings his tail down to meet hers. The openings of their cloacas, called vents, touch for a moment, sperm is transferred, and that is that.

So might this pair's liaison have ended had the female not seen a male nearby she thought more fetching. She scoots out from under the first male, who tumbles to the ground and lies in a befuddled heap. The second male hustles over to the female to commence circle dancing, flipper-patting, and copulation. Dee veritably cackles when they finish. "Look at you!" she says to the female, who was sharing a loving gaze of

Dee Boersma interacting
with one of her study
subjects.

post-copulative bliss with her new paramour as if they had been together for a hundred years. "You hussy! You little hussy!" Then she whirls away.

She can hardly wait to explore this place she already knows so well. She takes over the Gee Whiz notebook and fills it with terse prose in hurried script. We had been told to use the book for notable events—the first egg, an early visit from the provincial governor—but Dee's idea of what constitutes notable is far more expansive than ours.

October 8

Turbo went to sea and got clean and was looking to come in by 7:30 p.m. He had a short visit in the house standing next to Eric. Turbo looked like he wanted pancakes too. 7 days without H_2O is quite a feat. Turbo looks fatter and very soft.

Went and did top of road where there is a boardwalk. It's the turn in the old road and [illegible] but this is now a boardwalk. Within the fenced area were 95 nests with 157 penguins. I saw one egg and maybe 3 ♂ wander.

October 11
Very windy day. Clay and Dee locked in the Cueva for about an hour because the lock broke.
Turbo brought a piece of grass to the door this evening at about 8 p.m. and dropped it and went out the door. Turbo saw his reflection in the shiny brown paint of the door when he had his head bent in [a] circle dance and he clucked at it.

The days with her are exhausting, as Ginger warned us they would be. We are up earlier, we no longer linger over meals—we hadn't realized we were lingering—and we often find our way back to the house by the light of the moon. "This says I walked twelve miles," Dee says one night over dinner, fingering the pedometer she wears around her ankle to track her steps for the day. "Funny, didn't feel like twelve miles."

The energy of our work also changes. Before, we had been led by the project's detailed protocol, and Ginger was its chief adjudicator, nipping at our heels. Now that Dee is here, we respond more to our (or her) exuberance and curiosity. Anything we can see we can look at, and anything we can look at we can measure, so we might as well.

Almost every one of Dee's thoughts come from this accumulation of data. From it she has helped write over one hundred papers and however many book chapters, and edited a book or two. After watching her for a few days, I have come to consider this a mere fraction of her output. She holds another bibliography in her head, a vast body of gray literature: germs of ideas, stray notions about the finer details of penguin being. "I wonder whether chicks are more likely to hatch within twenty-four hours of when they hear their parents' mutual bray," she muses one afternoon. "When they first star the egg to when they pip it to when they hatch, is that a fixed time? Or will they slow down or hurry up depending on knowing their parents are switching off? . . ." Then the makings of a way to find out unspool as she walks along.

"Sometimes the things she will say sound a little out there," Ginger later tells me, "but she usually turns out to be right." In her tone I hear rueful admiration. She and Dee have an interesting relationship. Ginger likes structured inquiry. She was a computer programmer for several years before she went back to school to study copepods. Dee's approach, grabbing snatches of data, sometimes making things up as she goes along, or tossing aside decisions that everyone had worked out months ago in Seattle, can drive Ginger crazy. The reason is her allegiance to the Database. The Database is the lab's digital archive of every measurement ever made at Punta Tombo, all painstakingly entered from the scads of yellow field notebooks—like the ones we are currently filling—that line the lab's shelves.

Dee after sampling the contents of a squid egg case.

At heart, I think of the Database as Dee's answer to the acknowledged limits of her method. She is conscious that much of her work is based on observation, rather than manipulation (that is, experiment). She and Ginger have friendly but frequent debates about the relative merits of consistency. Dee argues that the field is hardly a controlled environment. We must be flexible, nimble, responsive to its whims, the questions that might pop up out of nowhere. This is why she wants us to record everything: you never know when or how it might be useful. "When I set out, I wanted to have a way to answer questions I didn't know I was going to ask," she says. It is due to this flexibility that heaps of studies are scattered like precious gems throughout the Database, waiting for someone with the necessary skill and patience to unearth them from three decades of measurements.

Yes, Ginger counters, but for that to be possible we need the data to be collected systematically. Only then can the data support the questions you want to ask and let you compare season to season. You can't just decide to do something, or not to do something, or to do something differently because it feels right at that moment. That's why we have a protocol. Otherwise it's just a bunch of numbers.

"I know, Ginj, I know," Dee says, standing at the sink after another go-round over lunch, but when she turns to wash her plate she is grinning a little, because she also knows she can't help herself, and this might be the deeper truth.

᠆᠊᠊

Between Dee and all the eggs, time races along, and before we know it, it is the end of October. She and Ginger will head back to Seattle tomorrow, leaving El and me in nominal charge of the project. A daunting thought.

On her last evening, Dee decides we should all go to The Point to count the cormorants. She likes to check in on them; she remembers when there were thousands, before the kelp gulls moved in and squeezed them out. She is telling us this and other tales while we walk along the cobble beach past the end of the tourist trail when she spies dozens of translucent jellies marooned on the rocks by the outgoing tide. "Look at this," she says, stopping mid-tale. The jellies are about the size and shape of hamburgers, with a ribbed fringe. She picks one up, sniffs it, and then—good lord—takes a bite out of it. Not a big bite, but not a small bite, either.

"Yech!" she says, spitting out the mouthful and tossing the jelly to the ground. "Not sure that was such a good idea!"

She resumes her foraging. Among the beached jellies she finds a mass of squid egg cases. Each slimy, ruddy beige case is about five inches long, with the tiny individual eggs swirling in murky liquid. Dee grabs the mass and holds it up and cocks her head. I wince and think, *She wouldn't, oh my god, no, please don't*, but she plucks off one of the cases as if it were a grape, puts it in her mouth, and sucks out the contents.

"Whew!" she exclaims, puckering her lips. "Salty!"

We continue on. If Dee has tried to teach us anything, it is that one must use all his or her senses to experience Punta Tombo. Thus far I have been content to limit myself to sight, hearing, smell, and touch. In spite of this performance, I see no reason to change that.

Soon, we reach the outer limits of the kelp gull colony. "Wow, this is really getting big," Dee says as gulls lift off to hector us. We are moving around the colony edge when something up ahead catches her attention. She stiffens like a pointer and walks quickly to an enormous rock in the middle of the gulls. She points gleefully. "Look at this guy!" she says. "What is he thinking?"

Wedged in a small opening at the corner of the rock is a single male penguin. The rock is massive, tabular. It looks like it fell out of the sky and pinned the penguin where he lies. I crouch down and peer at him as the gulls wail their distress. He doesn't even head-wag; instead, he stares at me with a desperate look. *Don't tell them I'm here!* his eyes seem to plead. I stare back, similarly disbelieving. What on earth could convince a penguin that nesting in a kelp gull colony is a good idea? Then another question: Who on earth would even think to look for a penguin here?

I get up to say something to the one person who would, but Dee has already moved on, drawn to other stimuli. Now she is far away, a lone figure in the pampas grass. Kelp gulls fret the sky above her, but she has her back to them and to us. She is looking west toward the colony, seeing whatever it is she sees when she looks at hundreds of thousands of penguins, thinking whatever it is she thinks, which must be unlike what anyone else sees or thinks when they look at hundreds of thousands of penguins.

She may be leaving tomorrow, but she will be back in December. She always returns. Punta Tombo is hers, and Punta Tombo is her. Earlier, I had asked if she thought she would ever hand the project off to anyone else, or at least to scale down her own involvement to a more leisurely remove, as would befit someone like her, a conservationist in winter. "I don't think I'll ever stop," she said curtly, giving me a sharp look. She had big plans and seemed put out that I would suggest she was mortal, just like everything else.

4
The Food Web

It is a warm night in November, a few days after Dee and Ginger have gone back to Seattle. Briana, Emily, El, and I are all sitting around the table in the kitchen, flipping through the pages in the Cañada books and writing the next day's date at the bottom of each. We do this now so we won't have to tomorrow morning when we are in the field, thus saving a few seconds as we breeze from nest to nest and harvest data.

Near the end of the book, I come to the page for nest C30A. I bring it to mind: C30A, a good nest, but not a great one, in a *lycium* bush forty yards from the edge of the Cañada. Its entrance faces east, so it catches the sun in the morning. Thin branches form a trellis over the nest cup, obscuring it, but not completely. I can stand over C30A and see a penguin within without much trouble.

I don't often dwell on individual nests like this. It is the paradox of the Cañada. We make our rounds through the area every day, like mail carriers or prison guards, and are so familiar with the nests that they have faded into the scenery. Most of the Cañada penguins are likewise habituated to us, head-wagging in a perfunctory way when they bother to at all. But here I am reminded that each nest is a thread in the season's larger weave, adding its own small story.

I look down the page, which is coated with a fine layer of grit. C30A began this year like most nests, with the zeroes of emptiness:

Date	Time	Contents
13 Sep	1210	Ø
14 Sep	900	Ø
15 Sep	850	Ø

16 Sep	1130	Ø
17 Sep	820	Ø

(Already the story has faded. Why were we so late on the 16th? I can't remember, and neither can El. September seems a lifetime ago.)

After a few days, the male arrived, a bird of unknown age with the band number 52937. He had used this nest for the last two seasons and duly laid claim to it and the page:

19 Sep	900	52937
20 Sep	840	52937
21 Sep	910	52937

52937 was alone for almost three weeks, until one morning we came to find that his mate had joined him. A bird also of unknown age, her band number was 52575. She carried a satellite tag a few years ago, and this makes her, as we say, a sat-valuable bird. With her arrival, C30A therefore got the relative prestige of orange and green flagging:

9 Oct	850	52937 + 52575
10 Oct	830	52937 + 52575
11 Oct	840	52937 + 52575

According to notes at the top of the page, 52937 and 52575 have been a pair for at least two years, although I don't know if they ever successfully raised chicks. In any case, they were together at C30A for nine days, standing or resting or mutual preening, as penguins do. Then, on October 18, 52575 laid her first egg:

18 Oct	830	52937 + 52575 + #1E

Over both penguins' objections, we pulled 52575 and weighed her. She weighed 3.55 kilograms, which is a little less than eight pounds—good, but not great. Her egg was 7.22 centimeters long, 5.45 centimeters wide, and weighed 121 grams—not too big, but not too small. It was cool to the touch, because 52575 wasn't incubating it, instead letting it lie in the dust behind her, probably right where she had laid it. Penguins aren't obliged to incubate their eggs every second of every day, but we were still concerned. Punta Tombo doesn't succor penguins that stray much from convention.

The next day, assured of his parentage as much as he could be, 52937 left to feed himself for the first time in more than a month. He would swim northeast, in search of the schools of anchovy and hake making their way down the coast from Brazil. If all went as planned, he would return in a couple of weeks to relieve 52575, who stayed behind with their egg:

$$19\ \text{Oct} \qquad 830 \qquad 52575_I\#1E$$

(We were encouraged that she was now incubating the egg, as shown by the subscript I.)

A couple of days later, 52575 laid her second egg:

$$22\ \text{Oct} \qquad 900 \qquad 52575_I\#1E + \#2E_{ØI}$$

This egg was 7.58 centimeters long, 5.46 centimeters wide, and weighed 122 grams. That it was larger than the first egg was unusual, but not overly so. That 52575 was not incubating it—hence the subscript ØI—was more vexing. Why is she so unconcerned about her eggs after she lays them? But the next time we see her she was sitting on both, just as she ought to be:

$$25\ \text{Oct} \qquad 800 \qquad 52575_I 2E$$

Penguin chicks take about forty days to hatch. As her mate had in September, so did 52575 settle in now. Every day we came to C30A, and every day we found her in more or less the same place and more or less the same pose. "There you are!" we would say. "You keep those eggs nice and warm!" Then we would sashay off to the next nest, leaving her alone to attend to her clutch, the days stretching out before her.

$$26\ \text{Oct} \qquad 830 \qquad 52575_I 2E$$
$$27\ \text{Oct} \qquad 850 \qquad 52575_I 2E$$
$$28\ \text{Oct} \qquad 800 \qquad 52575_I 2E$$
$$29\ \text{Oct} \qquad 900 \qquad 52575_I 2E$$
$$30\ \text{Oct} \qquad 820 \qquad 52575_I 2E$$

Over the past few weeks, I have started to think of incubating penguins as patience incarnate. They are so silent, so still, so stoic, or so it seems. Weeks ago, they had the entire ocean to range across if they so chose, but now their aquatic gifts have been rendered useless, are a burden, even, bound as the birds are to this square yard of dirt. The nest has become the seat of their being. In it, they place a simple animal faith: that it will keep their eggs safe and hidden from all the things that want to eat them. It is a lot to ask of a bush or a hole in the ground. Sometimes too much.

31 Oct	820	#1E (shell of m2E 3m NW—gull!)
1 Nov	750	#1E
2 Nov	800	Ø #1E in 3 large pieces in front—gull

What the codes say is that on October 31, we found 52575 gone and the #1 egg alone in the nest. Three meters to the northwest were the remains of the #2 egg, its contents eaten by what we assume was a kelp gull. The next day, the #1 egg was still there alone, but by the following morning a kelp gull had eaten it, too, dragging it to the nest entrance before hammering it to pieces.

The leftovers from a kelp gull's repast. The broken halves of eggshell are diagnostic of avian predation. If a peludo eats an egg, it leaves bits and pieces of shell scattered about like confetti.

52575 came back six days later. This was a couple of days ago. She was still at C30A this morning, just sitting there. It is hard to read a penguin's mood, but to me she looked like she didn't know quite what to do with herself.

9 Nov	830	52575
10 Nov	810	52575
11 Nov	850	52575

While we will never know exactly why 52575 left, we can guess. Based on her lightness when she laid her first egg, she probably got hungry and went to feed herself. We can hardly begrudge her that. To abandon her clutch for the sake of her own well-being was a rational decision. Penguins live a long time, and she will breed again, but we pass a judgment on her nonetheless, even if we don't call it that. We will no longer check C30A, no longer care, officially, what becomes of 52575 or 52937. For our purposes, their story has ended.

I remove all the pages for C30A from the binder and put them in a Ziploc bag we have for failed nests: those that were never occupied, those where the male never got a mate, or those, like this one, where the pair lost their eggs to a predator.

Four hundred thousand penguins can hardly gather in one place and expect to go unnoticed, but so far we haven't had to face predation in any broad or significant way. Occasionally we come across a penguin's savaged body—the work, Ginger told us, of a *colpeo* fox that roams about the colony—but such deaths are rare enough to seem extraordinary.

Eggs are another thing. For us, they are nascent penguins. For other animals, they are hearty mouthfuls of fat and protein, conveniently packaged and produced in bulk. The skunks and weasels and such, which before were minor curios in the Punta Tombo menagerie, have lately become more sinister. The hairy armadillos, called *peludos*, for instance. When I first saw them bumbling around, I liked to think of them as nearsighted little tanks. They bumble still but are no longer so comically harmless. The penguins lunge at them when they pass, hiss at them, hammer on their hard backs.

By far the most frequent eater of eggs are the kelp gulls. I have watched their rise to prominence with some ambivalence. Most of my dealings with kelp gulls have consisted of them screaming at me. "The voice of the Kelp Gull has a considerable range of expression," Robert Cushman Murphy has written, but I have yet to hear much evidence of this. *Ki-och! ki-och! ki-och!* is how the guide books translate their incessant call. It is just white noise to me. I grew up around gulls like these: the same size, the same ubiquity, the same squawking. I was happy to tune them out at home and have been happy to tune them out here.

Now they are harder to ignore. Daily we consign nest pages to the Ziplocs of Failure, as pair after pair of penguins lose their eggs. When we go to check Cañada, we see dozens of gulls perched on the handrails of the two short pedestrian bridges along the tourist trail, searching for their morning repast. More float over the colony. Others traipse about on the ground as if window shopping. When one sees an unattended nest, or even a penguin with its back turned, it hurries over to try to snatch an egg. It might bash its prize open right in front of the penguin. (Since the egg is no longer in the nest, the penguin doesn't necessarily know to care about it.) Or the gull might take the egg and fly up and drop it to the ground, fluttering down to lap up the smear. More than once I have found an egg wedged in the branches of a *molle* or *duraznillo* bush, cold and dead and dirty, but otherwise intact. Dropped and forgotten.

If I have set up the penguins as birds of faith and their nests as domestic temples, then are those creatures that violate the temples necessarily profane? When she was here, Ginger would caution us to avoid any undue prejudice against predators. "Remember, all the other animals are trying to feed their young, too," she would say. "Don't the gull babies deserve to eat?" I concede grudgingly that they do, but my distaste for kelp gulls has other layers. The hunger of their search is aggravating, the way they check every single nest to see if it contains something of interest. They have a researcher's vigilance and attention to detail not so different from my own.

In clearer moments, I know I'm being unfair. I tell myself the kelp gull is a beautiful bird. I even murmur it under my breath: *the kelp gull is a beautiful bird, the kelp gull is a beautiful bird.* As a mantra, it has the benefit of being true. Large and clean-lined, ivory white, back black as carbon, and a skilled flier, the kelp gull is a beautiful bird. They are soothing to watch as they ride the wind, matching its gusts, before tilting their bodies and wings just so and sailing fifty yards to some point as if drawn there on string. A good exercise, to consider the kelp gull outside of its role as predator, ne'er-do-well, unflattering mirror—outside of the fact

An adult kelp gull.

that, whenever I see one steal away with a penguin's egg, I feel a hot flare of hate.

⌒

Nothing has been at Punta Tombo forever. Kelp gulls were first recorded here in 1876, when a British ornithologist named Henry Durnford saw fifty pairs on The Point. Those numbers have since climbed. In the last official census, completed a couple of years ago, some of Dee's Argentine colleagues counted 6,457 pairs. The colony is now one of the largest in Patagonia, and Dee has watched it grow with a mix of awe and alarm. "They're just eating up the land out there," she said during one of our walks. When she left, she asked that we count all the gull nests the next time we have a free afternoon.

Free afternoons are rare, and a couple of weeks pass before Briana, Emily, El, and I can finally tramp out to The Point. The sky is a dull, soiled gray; rain spittles down. The closer we get to the gulls, the more potent the smell of ammonia. When we arrive at the colony's near edge, we confront a solid mass of birds spread as far as we can see. Their heads are arrayed like row upon row of white pushpins, and they all seem to be staring at us. We look at them, then at each other. Somehow, the four of us are supposed to wade into their midst and not only count every single one of their nests, but also keep straight which nests we have already counted and which we have not, as well as those which someone else might already have counted and those which someone else might already have not.

I feel our enterprising spirit wane somewhat. "How do we want to do this?" I ask.

No one says anything. The first phalanx of gulls rises and starts to whirl over our heads. Their screaming makes it hard to think. Small flecks of white mix with the rain spotting my jacket. Gull poop. Great.

Emily, who spent last summer at a tern colony in Maine, says, "With the terns, we used popsicle sticks to mark the nests."

A fine idea, but no one has thought to bring ten thousand popsicle sticks.

Briana says, "We could just fan out about a hundred meters apart and try to count in lanes." She shrugs. "Unless someone has another suggestion."

From such decisiveness do the data come. We space ourselves accordingly and begin a careful trudge forward. The best way to count the nests, I find, is to let my eyes sweep back and forth across my lane, while *click*ing at an appropriately brisk rate. Under the umbrella of screeching gulls, I fall into a trance of census. My steps become slow, rhythmic, my head sways as if only loosely attached to my body, and my thumb is a distant, perplexing appendage, madly working the clicker button as nests pass across my visual plane.

Clickclickclickclickclickclickclickclickclickclickclickclickclickclickclickclickclickclickclick

At Punta Tombo, kelp gulls don't lay their eggs until early November. They will continue to lay through late December. As I dodder along, those gulls at the nests nearest me stand as if coming to attention. They shuffle in place and natter worriedly (*kek kek kek!*), the powerful urge to flee vying with the similarly powerful urge to protect their eggs. Some stand fast, some back away, some leap into the air and join the wailing throngs. A few of the nests they leave behind have a full complement of three eggs, but most have only one or two. Huddled against one another in the nest cup, the eggs are long and conical. They range in color from dark beige to light blue, and are speckled brown or black, as if dabbed with varying intensities of gouache.

Clickclickclickclickclickclickclickclickclickclickclickclickclickclickclickclickclickclick

As camouflage, it is quite lovely.

Clickclickclickclickclickclickclickclickclickclickclickclickclickclickclickclickclickclick

Kelp gulls are the widest ranging seabird in South America, if not the entire southern hemisphere (or even, as Murphy suspected, "among sea birds"). In addition to both coasts of the continent, they breed in southern Africa, Australia, New Zealand, and across the sub-Antarctic islands. A few have even strayed to North America, where they hybridize with herring gulls in Louisiana. In Argentina, they are found almost everywhere, not only along the more than two thousand miles of coastline between Buenos Aires and Tierra del Fuego, but also hundreds of miles inland, at wetlands and lakes and sewage ponds. They are by far the most common gull in Patagonia, their numbers increasing by almost 3 percent per year.

Clickclickclickclickclickclickclickclickclickclickclickclickclickclickclickclickclickclick

Like most seabirds, kelp gulls prefer to nest on islands. More than 80 percent of their colonies in Argentina are offshore, and the farther from the mainland, the better. To risk breeding on the mainland with all of its predators, they must have a compelling reason, as these gulls clearly do. (Penguins.) Elsewhere, new colonies continue to sprout and expand.

Clickclickclickclickclickclickclickclickclickclickclickclickclickclickclickclickclickclick

Among the many reasons kelp gulls are so common and widespread is that they will eat almost anything. Their creativity and enthusiasm for experiment tests my tepid resolve to see them in a more charitable light. I have read, for example, that since the 1970s, kelp gulls from the Península Valdés will land on the backs of southern right whales and rip out chunks of their flesh. They focus on mothers and calves, which use the Golfo Nuevo as a nursery. In response, the mothers have changed their swimming and resting habits. They arch their backs to keep them underwater, leaving only their heads and tails exposed, sparing them attack. The technique, called galleoning, is born of years of experience. Researchers first noted the behavior in 1984, and it has since spread among the mothers. But calves, being young, don't know to avoid the gulls. Almost every single one of them has ugly lesions on its back as a consequence—a grim visual record of youthful naivety.

Clickclickclickclickclickclickclickclickclickclickclickclickclickclickclickclickclickclickclickclick

Between 2003 and 2014, 626 right whale calves died at Peninsula Valdés—an average of more than 50 deaths per year, and a sharp increase from the decade prior, when only 82 died. No one is sure why so many more calves are dying now, and while scientists don't hold gulls responsible, they also say the gulls certainly aren't helping.

Clickclickclickclickclickclickclickclickclickclickclickclickclickclickclickclickclickclickclickclick

Some whale advocates want to shoot the problem birds, as they call those gulls that harass the calves the most. This mix of management and revenge is all too human a response, but while the whale-gouging behavior is spreading among the gulls at Valdés, it is good to remember that only a few individuals practice this particular, peculiar art. The same could be said about gulls and penguin eggs at Punta Tombo. I may imagine a generic kelp gull trying to steal all the Magellanic penguin eggs, but kelp gulls here mostly eat fish and marine invertebrates: crustaceans, such as crabs; and mollusks, like small mussels and snails, which they tear off the rocks. A small fraction of gulls eats seabird eggs or chicks (although Dee has found that, when a penguin deserts its nest, a kelp gull usually finds the eggs first). Gulls are about as likely to eat our garbage as they are penguin eggs.

Clickclickclickclickclickclickclickclickclickclickclickclickclickclickclickclickclickclickclickclick

Problem birds. As the kelp gull's population in Argentina has soared, the consensus among scientists seems to be that more of them is a bad thing. There is a long list of reasons for this: kelp gulls eat other birds' eggs, they compete for space with less common species (such as the cormorants on The Point), they eat the offspring of animals more generically appealing than they are (penguins, right whales). They can also smash into planes at airports, they might be a threat to human health in some way, and so on. Such accounting serves to distract us from the kelp gull's truest ingenuity. More than any other seabird, kelp gulls have learned to take advantage of the world we have made for them. Human and kelp gull are coupled. We expand, they expand with us. Their mere presence has now become, in the words of one biologist, a sign of a degraded environment.

Clickclickclickclickclickclickclickclickclickclickclickclickclickclickclickclickclickclickclick

I can only assume this person means a peopled environment. I wonder if he is being purposefully indirect, masking a misanthropic critique in ornithological code. The better question might be whether it is possible for an animal to be just an animal anymore. The kelp gull is no longer just a kelp gull. It has become an argument.

Clickclickclickclickclickclickclickclickclickclickclickclickclickclickclickclickclickclickclick

Or, in this case, an indictment.

Clickclickclickclickclickclickclickclickclickclickclickclickclickclickclickclickclickclickclick

Throughout southern Patagonia, the kelp gulls' greatest boon has been the fleets of trawlers at sea, and on land the garbage dumps and fish processing plants around Puerto Madryn and Rawson. These plants produce on average about fifty thousand metric tons of fisheries waste per year, although the amount can vary. Gulls swarm the plants by the thousands, by the tens of thousands. They gorge on the parts of a fish for which we have no appetite: bones, fins, bloated heads, bits of filet. Biologists estimate that fishery waste alone could support more than a quarter of a million kelp gulls around here.

Clickclickclickclickclickclickclickclickclickclickclickclickclickclickclickclickclickclickclick

Per the protocol, every five days we are to walk along the tourist trail at night and count the flood lights of the trawlers working within sight of Punta Tombo. Those boats—the source of much of the fishery waste—are easy enough to see, even though they are miles away. Their lights dot the horizon, as many as nineteen of them, bright orange stars in a flat constellation: the Fishing Line.

Clickclickclickclickclickclickclickclickclickclickclickclickclickclickclickclickclickclickclick

The trawlers are fishing for Argentine hake, a codlike species that can grow up to three feet long and weigh about eleven pounds. Schools of hake are found in deeper waters, about one hundred fifty feet or more beneath the surface. Kelp gulls can't dive for their prey, so when the trawlers haul the hake up and throw the ones that are below regulation size back, they make food available that gulls wouldn't be able to access on their own. The amount of bycatch and discards tossed from the trawlers is so great that the gulls can be picky, choosing only the most perfectly sized morsels of their desired fish, the hake, and leaving the rest.

Clickclickclickclickclickclickclickclickclickclickclickclickclickclickclickclickclickclick

Sometimes when I look at the orange stars through my binoculars, they have a fuzziness about them—an electron cloud of shrieking kelp gulls.

Clickclickclickclickclickclickclickclickclickclickclickclickclickclickclickclickclickclick

I am familiar with the effect.

Clickclickclickclickclickclickclickclickclickclickclickclickclickclickclickclickclickclick

Magellanic penguins also eat hake, although where the trawlers pursue the adults, they pursue the juveniles. Diet analyses of the penguins from Punta Tombo and nearby colonies show that, more than hake, they prefer the Argentine anchovy, or *anchoita*, a smaller fish that forms dense schools. But like gulls they are opportunistic feeders, if more restricted in their opportunism since they feed only at sea. Unlike gulls, they can dive for their prey. Sometimes they dive as deep as three hundred feet, but usually they stay around ninety feet or so. (During the summers, anchovy tend to gather between ninety and three hundred feet deep; they swim deeper in the winter.)

Clickclickclickclickclickclickclickclickclickclickclickclickclickclickclickclickclickclick

Some penguins are not above following the trawlers in the hope of an easy bite. Underwater, they zip in to snatch fish from the nets as the boats haul them up.

Clickclickclickclickclickclickclickclickclickclickclickclickclickclickclickclickclickclick

The bite is not always so easy. Penguins can get caught in the nets and drown.

Clickclickclickclickclickclickclickclickclickclickclickclickclickclickclickclickclickclick

In 2006, Pablo Yorio, a biologist with Centro Nacional Patagónico and one of Dee's former students (or, as she likes to call them, her chickies), coauthored a study that examined how seabirds use the trawl fishery's discards in Golfo San Jorge, a large bay about one hundred miles south of Punta Tombo. Penguins were less likely to benefit from bycatch and discarded hake than gulls were, but they were more likely to die for their efforts. During a two-month period, Pablo estimated that more than fifteen hundred penguins were caught in the nets and drowned; three hundred imperial cormorants also died. Both totals, the authors noted, were likely underestimates. Many more penguins were probably killed, in part because hake trawlers operate closer to their breeding colonies than boats from other fisheries.

Clickclickclickclickclickclickclickclickclickclickclickclickclickclickclickclick

The way penguins and gulls interact with the hake trawlers in southern Argentina recapitulates a global dynamic. Penguins, since they dive for their food, are members of the group of seabirds called pursuit divers. This group, or guild, also includes murres and puffins in the northern hemisphere. Where fisheries are concerned, kelp gulls are considered scavengers, along with birds like albatrosses and petrels. A third guild called kleptoparasites, which includes skuas, harasses gulls and other scavengers until they drop their meals.

Clickclickclickclickclickclickclickclickclickclickclickclickclickclickclickclick

When people talk of fisheries and seabirds and death more generally, scavengers are usually thought to suffer the greatest mortality, but when the effects are parsed by the major types of fishing gear—the longline, the gillnet, and the trawl—a different pattern emerges. In longline fisheries, where a single line with hundreds of baited hooks is set in the open ocean to catch tuna or halibut or swordfish, it is true that scavengers die most frequently. Hundreds of thousands of albatrosses, petrels, and shearwaters are drowned all over the world when they try to snap the bait from the hooks and get dragged underwater. But Dee has argued that with the other two gear types—the gillnet and the trawl—pursuit divers fare worse.

Clickclickclickclickclickclickclickclickclickclickclickclickclickclickclickclick

Gillnets are fine mesh nets that can be set anywhere in the water column. They hang in the ocean and are effectively invisible. Gillnet fishers and seabirds often pursue the same prey, and the diving birds are caught by the hundreds of thousands, entangled when they swim into nets they cannot see.

Clickclickclickclickclickclickclickclickclickclickclickclickclickclickclickclickclickclick

Trawlers catch fish by towing large, weighted nets behind them. The nets either drift midway through the water column or scrape over the sea bottom, and once full of quarry are hauled back onboard with heavy cables. Trawls are thought to be one of the more benign gear types; flying birds sometimes smack into the cables, but studies have found several ways to prevent injury, such as by hanging flags or streamers. The nets, though, can ensnare diving birds as efficiently as they do everything else. Meanwhile, bobbing along at the surface, the scavenging birds blithely feast on the discards that get shoveled from the boats.

Clickclickclickclickclickclickclickclickclickclickclickclickclickclickclickclickclickclick

Such is the case in the Argentine hake fishery. With the trawlers, the gulls profit the most from the food subsidy the boats provide, and are only rarely injured or killed. Magellanic penguins, on the other hand, not only compete directly with trawlers for food and are occasionally killed by them, but they also suffer indirectly when the gull's population grows, in part as a result of this reliable food source. The surfeit of gulls turns its predatory attention to the penguins back at the colony.

Clickclickclickclickclickclickclickclickclickclickclickclickclickclickclickclickclickclick

In the end, kelp gulls are helped on balance, while the penguins are doubly or even triply harmed.

Clickclickclickclickclickclickclickclickclickclickclickclickclickclickclickclickclickclick

I reach The Point, and, blessedly, the last smattering of kelp gull nests before the bare strip of no-bird's-land that separates them from the beleaguered cormorants. Every rock is painted over with a thick layer of guano. The smell is penetrating, but mixed with the scents of the sea, not entirely unpleasant. I breathe in, hold the air in my lungs, and add a few more *clicks* in case I missed a nest or two along the way. El finishes her lane a couple of minutes after me, and then Briana and Emily. All of us look a little stunned as we regroup. When we tally our clickers, we learn we have counted a total of 8,789 nests. Seeing this number, we grin at each other. It feels oddly satisfying to have met the kelp gulls' teeming absurdity on the field of quantification and triumphed over it with equally absurd precision.

We turn to walk back. I watch my feet as I go—I don't want to crush any of the gull eggs—and that is when I get my first close look at the nests that my eyes had only slid numbly over before. Some are made mostly of shredded pieces of red algae, others with sheaves of green. A few have bits of brittle pink coralline algae included as accents of color. Some nests have no marine materials at all, but are instead made from terrestrial objects, with pampas grass and weathered sticks laid in swirls, the former stacked like a warm bed of straw, the latter artfully gnarled. Some nests have feathers or pebbles or shell as the base. The eggs in these are delicately poised on the unstable rubble; those resting among the beds of algae seem more cushioned. I wonder if this reflects competing schools of kelp gull parenting philosophy, whether the birds that balance their eggs on shells expect different things from their chicks than those that bed their eggs in seaweed. (Probably not.) Some nests have one or two bones among the softer materials. Here is one with a large primary feather artfully placed across the clutch like a fountain pen, pointing east. Here is one that veritably overflows with green algae and looks so comfortable I am tempted to rest my head next to the eggs and take a nap.

I slow to a crawl, then stop altogether, turning in slow circles. The cloud of kelp gulls descends. I am no longer beset only from above, but from the side, from behind, from all directions in this maelstrom of birds. They can't drive me away so easily. Here is another nest with its own subtle fineries, here another, here another. I would like to look at them all day, admire them, make a study of the full range of kelp gull aesthetic sensibilities.

Not much kelp, I can't help but notice. How delightfully contrarian!

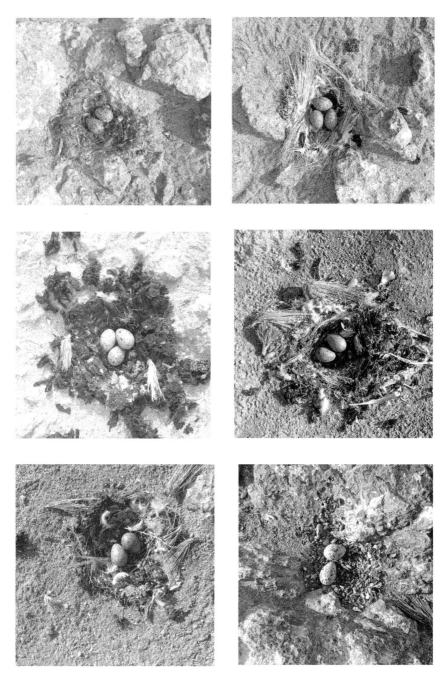

The material diversity of kelp gull nests.

5
Hatch

Two days ago the egg started peeping, a sound so near to silence that we almost missed it. "Wait," El had said. "Do you hear something?"

I hadn't. It was nine o'clock in the morning but the day was already hot. We were at 141G, a nest in Max-Vista. The area marks the colony's inland boundary, and the name Dee gave it reflects this. Half a mile from the beach, it is about as far as a penguin is willing to waddle to a nest (Max), and it sits atop a hill, affording a generous view of the bustle below and the sea beyond (Vista).

El motioned me down. I knelt next to her and leaned as close to the nest entrance as I dared. The female was coiled back, sure to strike if I moved in an inch. But yes, I could hear it: steady and insistent and muffled by eggshell.

Peep peep peep peep peep peep peep peep peep peep peep

"Wow," I said. El smiled. I made a note on the nest page and we continued on.

Yesterday, the egg was cracked (or pipped), the crack (or pip) about half an inch long. Loose bits of fractured shell wiggled as the chick strained to push itself free. We could see the tip of its small bill, the white egg tooth raised like a horn. Its movements were slow but determined, suggesting tenacity.

Today, the female's stance is changed. Before, she looked settled on her eggs, focusing all her warmth on them, but now she is hunched, as if straddling something. I prod her chest with my *gancho* until she backs away. Underneath her is a tiny, bedraggled chick, newly hatched and exhausted from the effort. It looks like a scrap of gray velour.

"Wow," I say again. I kneel and reach for my egg cup.

El says, "That is the cutest thing I've ever seen." She looks right at me. "You be careful with it, Eric Wagner."

There have so far been no fixed points to the season at Punta Tombo. Change instead comes in waves, one phenomenon rushing in as another recedes, until we realize that the earlier one, which had been so engrossing, our whole world, is over. The days of the single males, the arrival of the females, the laying of the eggs—these have become discrete periods only in retrospect.

A chick is definitive the moment it appears. Or even before that, from the moment it could appear. Here we take our cue from Dee. She may allow for a forty-eight-hour window around a new egg, but she insists on knowing the exact day the chick hatches from that egg. That is why,

An adult penguin stands over its clutch of two eggs. One of the eggs has a large pip; the tip of the emerging chick's bill is visible.

thirty-six days after the female in 141G laid her first egg, we started visiting the nest daily, on what are called chick-checks.

The first day, the male, band number 45787, was sitting on the clutch. His mate, who is unbanded, relieved him two days later. Watching them come and go, I again see how precisely calibrated a penguin's life has to be. In the Cañada, those pairs still with eggs have been switching off at their nests more and more frequently, their absences progressively shorter. When the males returned from that first long foraging trip, the females left to feed themselves, maybe for a week, or at most a week and a half. When they came back, the males went, but for only a few days. Then the females took their turn, also for a few days. Some mates have since been trading places every other day. They are trying to time it so one of them is at the nest with a bellyful of fish when the first chick hatches, and they can feed it as soon as possible.

Sometimes we are around when a female or male returns to its mate. Their wet feathers shining like obsidian, they waddle up to their nest, stick their head in, and bray. Dee says it's like they're calling out, "Honey, I'm hooooooome!" The nest-bound mate emerges and responds in kind. The two penguins will then stand in front of the nest and call together, their voices overlapping in a burry exchange of staccato trills and wails: *b-b-b-bbbbb-bwaaaa! (bbbbb-bwaaaa!) b-b-b-bbbbb-bwaaaa! (bbbbb-bwaaaa!)* They might do this for several minutes. If it is the male who has returned, he will occasionally bring a gift: tufts of grass he has yanked from the ground or a pebble that attracted his eye. ("In so far as these stones have strong or unusual colors," Robert Cushman Murphy noted, "they may signify a primitive aesthetic sense . . .")

The penguins' ritual of return is called a nest relief ceremony, and the act of the male and female braying together, a mutual display. Alan Clark, another former student of Dee's, wrote his dissertation on these duets. Having sung on and off Broadway for more than a decade before earning his PhD, he was drawn to the penguins' strange music. He spent several seasons crouched behind bushes in the campo, his shotgun microphone pointed at pairs so he could record their natterings. (To muffle the wind, he wrapped the microphone in a fuzzy toilet cover.)

Like other birds, penguins make their sounds with what is called a syrinx, from the Greek word for pan pipes. Unlike the human larynx, which sits atop the trachea, a penguin's syrinx is at the bottom, where the trachea forks into the lungs. An air sac surrounds the structure of membrane and bone, and it lies deep in the chest cavity, which acts as a reverberating chamber. The syrinx has two sides that can vibrate independently, creating what biologists have called the "two-voice" phenomenon. This explains the eerie polyphony of the penguin's call: the penguin is making two sounds at once. But where I hear their calls as a hodgepodge of huffs and howls, Alan heard more than twenty discrete vocal structures: overtones, harmonics, other resonances that encoded information about the caller. Smaller penguins sound different than larger penguins. ("Higher pitched and whiny," Alan says.) Older penguins sound different than younger penguins. ("Hoarse, like they're chain smokers," Alan says.)

Alan then played the penguins' calls back to them and their mates. He found that, for Magellanic penguins, voice is identity. They don't distinguish each other by sight alone, but also by sound. Both males and females get excited when they hear their own mutural display calls, but not those of other pairs. Also, a female perks up when she hears her mate's ecstatic display but ignores those of both neighbors and strangers. Curiously, or perhaps not, the response cools the longer the male and female have been together as a pair. ("Draw your own conclusions," Alan says.)

Alan showed something else as well: a chick will respond to the mutual display of its parents. While in the egg, it was listening, learning their calls. Once it hatches, it has already started to know their voices.

⌒

I heft the egg cup and prop myself on my elbows to see how best to remove the chick. Although it is the first time I've had to do this on my own, this isn't the first study chick of the season. One hatched a few days ago in the Factura area, at a nest in a *molle* bush on the berm. Two field workers from last year, Jeff and Olivia, processed it. They flew down from Seattle expressly to teach us how to handle chicks, how to measure them. They made it seem simple, their movements skilled and sure, but the lesson ended on a note of caution. "Working with chicks is a lot different than working with eggs," Jeff told us. "You'll see that for yourselves." Yes, we will. An egg might break, but a chick can die.

I scoot the egg cup into the nest and set it next to the female. Thankfully she is calm, nibbling on the egg cup rather than battering it. I cover the chick with the cup and wedge my *gancho* under its belly to hold it in place, and swiftly withdraw the assemblage. The female watches all of this with an almost academic curiosity. El takes the chick in her hands, and I drop the egg cup to the ground. My arms suddenly feel boneless. I didn't realize I was so tense.

El holds the chick, which seems not to be touching her skin so much as floating on a cushion of her care. Its feet are fleshy and pink, its flippers soft and pliable as felt. Its eyes aren't yet open. Some substance from the eggshell has dried to a crust on its head, so that tufts of down stick up like a mohawk. The chick peeps and nestles itself into El's hands. "Oh, my goodness," she breathes. From her vest pockets she removes a 100-gram scale, a pantyhose liner sock, and her calipers. She puts the chick in the sock and clips it to the scale. "One hundred grams exactly," she says as the chick dangles in the air. New chicks are supposed to weigh between seventy and ninety grams, which is about three ounces, so this one is notably heavy. El takes the chick out and turns it around in her fingers, pressing it lightly. "Feel how full its belly is," she says. "I think it was recently fed."

She next arranges the chick on her knee to measure its flipper and foot and its bill's length and depth, spreading out the relevant body parts between her forefinger and thumb. Its flipper is 2.72 centimeters long, and its foot 3.3 centimeters. Its bill is 1.54 centimeters long, but when El tightens the calipers to determine the bill's depth, the chick squeaks in distress. "Oh no!" El exclaims. "Sorry!" She hurriedly finishes. "Zero point eight three centimeters," she says, before adding sheepishly: "That might be a little bigger than the bill actually is."

Now that El is done, we each take a minute to hold the chick. We are almost giddy. No, not almost: we are giddy. I cradle the chick in my hands, nuzzle it, inhale deeply. It has a musky scent, maybe a touch un-guent. The smell is completely new to me. "Hello," I whisper. The chick is exquisitely soft and delicate. A penguin that can't see, and so doesn't have the wherewithal to fear me, or even to realize I am holding it. A breeze plays across its down. It yawns. Its little pink tongue has denticle buds. "Hello," I whisper again. The response is involuntary. I wonder what my voice sounds like, whether I even register as a thing, or if I am just ambi-ent noise, like the wind rustling a bush.

I put the chick in the egg cup and deposit it near its mother. She tucks it under her brood patch. I am getting up to leave when El says, "Let's give it a name."

"Okay," I say. "What should we name it?"

"The nest ends in G," El says. She thinks for a second. "George Xavier." It is the name of a good friend's young son.

"Sounds good," I say, and I scribble *George Xavier* next to the chick's measurements.

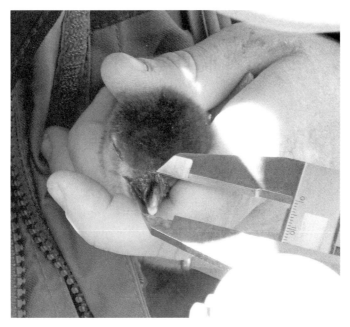

The somewhat fraught proposition of measuring a penguin chick's tiny bill with calipers.

⌐⟳

Until now, there has been a namelessness between us and the penguins. Outside of Turbo, the project carefully rations identity. Most penguins are simply male or female to us, or at a glance just birds. ("There was an angry bird on the tourist trail today.") The study penguins we know by their five-digit band numbers, which itself is a kind of numerical anonymity. ("52575 lost her eggs." "Oh, that's too bad.") A few penguins have actual, given names, but those were chosen by donors who paid a few thousand dollars apiece to sponsor them: the beloved Packy, Thorne, Caleb, Peach, TimKat, Wadsworth, Tweedie Two, and so on. We are to pay special attention to these VIP birds, as Dee calls them, so we can report on them to their benefactors. ("Peach was at her nest with a piece of grass on her head!" reads one representative note.) But their names are received and, outside of their financial significance, meaningless to us.

We name every single one of the chicks. We were advised not to. "It just isn't a good idea," Jeff said when we told him about George Xavier. We considered his warning but dismissed it in favor of our immediate delight. There will be Jetpack and Jambalaya in Bun 413J, Emerson and Emile in Fac 102E, Kumquat and Quiet in Bun 414Q. (Q nests sometimes messed

A newly hatched chick fits easily in the hand.

with our conceit.) And why not? Chicks are the softest thing we have seen in this otherwise hard land. Why should we not draw them close to us, instead of holding them at arm's length like we do their parents? Why should we not magnify the small?

Chick siblings rest after being measured.

6
Storm

Since the first chicks hatched a couple of weeks ago, the colony has been filling with them. To a chick they are all precious, tiny and soft, but handling them is more taxing than I thought it would be. You arrive at a nest and see from the eggshells cast to the side that a new chick awaits, ensconced somewhere beneath its parent. Again you drop to your knees, the egg cup clenched in your fist. Again you find yourself nose to bill with an angry penguin. Again you struggle to retrieve the offspring it guards, as sweat worms its way down the grime on your face. Some days, after hours of this on the baking campo, I wish there were no more chicks, so I could escape the grinding stress of them and go to the house and sit in the dark. Maybe in the back room under a bunk bed, like Turbo sometimes does.

El tells me I complain too much when I mutter something to this effect one afternoon in early December. She loves to cuddle the chicks. She rubs each and every one of them against her cheeks, talks to them, lets them burrow in her pockets.

We are on our way back from Rawson, the provincial capital, where we have gone to renew our visas so we can stay for three more months. (Getting a scientist's visa is more trouble than it's worth, so as one of the project's several idiosyncrasies we are here, technically, as tourists.) A long day to propitiate the whims of bureaucracy has brought out a peevishness in which I sometimes indulge. The municipal sky and an unhelpful government official had the same bright, false cheer, but as we get closer to the colony, the clouds turn dark and mean.

"Do you think it will rain?" El asks.

The *guardafauna* went through several flags during the season, as the strong winds relentlessly chewed them down thread by thread.

"No," I say. I don't understand why, but something about the topography of Punta Tombo seems to repel the worst weather. Tremendous cells have borne down on us, only to veer away at the last second and leave the landscape dry. But when we pull up to the house, the wind is whining through the guy-wires that hold up the tall radio antenna next to the *cueva*, and the fraying Argentine flag outside the ticket booth blows out so flat and straight it looks starched.

"Do you think it's going to rain?" El asks again later, as we climb into the trailer for the night.

"No," I say, but this time I am less sure as I pull the door shut.

⌣

Wind has been our constant companion at Punta Tombo, in Patagonia. This is a land famous for its winds, where a man once fired a gun into the gusts and was killed when the bullet was blown back into him. Or so the story goes. I'm inclined to believe it. The wind this night is more monstrous than any I've ever heard. It roars overhead, so big and powerful that it sounds like it has palpable mass. Then the rain starts, slowly at first, but building in force until it pounds the roof. The trailer rocks and shudders with the storm's fury. Deep in my sleeping bag, I feel all too well that I am a small thing huddled at the bottom of a great sky.

The storm rages all night. Although the worst of it has passed by morning, the sky is still a nauseated green when I poke my head out the trailer door. A few penguins are braying, but the mist muffles their calls, while the wind smells of newly rewetted guano. El and I squish to the house. Inside, Briana and Emily are enjoying a relaxed breakfast. (There is no point in doing the morning Cañada check until the rain stops.) Apparently, they inform us, the roof leaks. Water trickles down the walls, pools on the floor, meanders over to the door. We flail at it with a mop, but it seems to be showing itself out, so we decide to see how the penguins are faring.

Outside, the storm has knocked the colony's arid character askew. The ground at Punta Tombo is hard, an all but impermeable clay. It doesn't welcome water, and large puddles have formed on the tourist trail, some so big they look like inland seas. The once dry *cañada* of Cañada runs with shallow but swift rapids. Small groups of penguins stand on its banks. They bob their heads and look from the water to each other, back to the water. Their world was once so neatly demarcated: the water there, the land here, the two only meeting at the shore. With water in places it normally is not, and in racing abundance, they don't seem to know what to do. Some founder in the puddles, unable to decide whether to swim or walk. At the *cañada*, a pair tumbles into the torrent when the water suddenly chews the banks out from under their feet. They pop to the surface, bemused, and are swept around a bend.

The rest of the penguins are covered in mud. They look so morose that I worry they will catch their deaths from cold, but then I remember they are penguins and so are used to being wet. Prefer it, actually.

⌐

We wander around until the weather has calmed enough for us to do the Cañada check. Given how vulnerable I felt last night, even sheltered and dry and human, I approach the area filled with foreboding. Eighty chicks have hatched there, and most are still small. What a relief it is to find that everyone in the first few nests is okay, if a little damp. The feeling lasts until I come to a nest midway through the book, 819D. Yesterday, the pair had two chicks. Today, the female, band number 53172, is brooding her second chick, but her first is sprawled in the dirt in front of her, dead.

Per the protocol, whenever a penguin dies, be it adult or chick, we try to figure out what killed it. I retrieve the chick's body. It is cold and limp, and slick with dubious fluids. Mud cakes its eyes and bill; 53172 probably trod on it after it was dead. I shiver with revulsion, but death is still data, so I get out my scale and calipers. The poor thing weighs 182 grams. Its bill is 1.56 centimeters long and 0.81 centimeters deep. (I can be sure of this since I don't have to worry about squeezing too tight anymore.) Its flipper is 2.93 centimeters long, and its foot 3.27 centimeters. It is fat and fleshy, and to all appearances healthy, aside from being dead. Clearly it died of exposure during the storm.

"Sorry, little one," I say, and lay the chick in front of its nest. I stand too quickly, wobble a little. Two nests later, a female is brooding both her chicks, the first of which is living, and the second that is not. It, too, is wet, and died during the storm. In the nest after, an adult male sits on his two dead chicks. They had wedged themselves under him as far as they could, but their rumps were still exposed, and the rain soaked them, and the wind blew, and that was that.

A chick lying dead in its nest.

Feeling glum, I walk to the other side of Cañada, where El has been checking another group of nests. She is downcast when I reach her. She has so far had five dead or missing chicks. "When I heard the rain last night, I thought it sounded nice," she says as she holds a sixth, and lays it next to its sibling, the seventh. She looks at the body, touches it. "I didn't know it would be like this."

The storm could hardly have struck at a more lethal time. A small chick cannot regulate its own temperature. For the first three weeks of its life, it depends on its parents to stay warm. After that, its down will insulate it against the cold, but only if the down is dry. If it is wet and the wind blows against it, it becomes worthless. So we see the storm's most lasting effect: of the eighty chicks in Cañada, it claimed thirteen. Almost all of them were around ten days old: a little too big for their parents to cover them completely, a little too small to keep themselves alive.

I weigh and measure El's last two dead chicks while she sits next me, hugging her knees. After we're finished, we go back to the house and check a plastic rain gauge attached to a fencepost. It is usually full of just dead bugs, but today their carcasses float in 11.4 millimeters of rain. I have to check twice to be sure; yes, 11.4 millimeters. That is a little more than one centimeter, which is not quite half an inch. I think of the sodden penguins, the enormous puddles, the new river in the *cañada*, the dead chicks we left out in the open so the kelp gulls could find them. Half an inch of rain caused all of this. Half an inch!

꒰ꔛ꒱

When we speak of penguins and climate change, we tend to look first to the species that live in the colder parts of the world. The emperor and the Adélie, perhaps the two most well-known penguins, both breed only in Antarctica, and the poles are the fastest-warming regions on the planet. In the past few decades, mean winter air temperatures in Antarctica have risen by almost 11°F. One of many consequences is a shift in sea ice coverage, either up or down depending on a suite of factors. This is a threat to both penguin species. In some places where sea ice cover is increasing, emperors and Adélies are forced to waddle more than fifty miles from their colonies to open water—far too great a distance for chicks that might be waiting for their parents to return with food. In other parts of Antarctica the sea ice breaks up earlier or has disappeared altogether, either sweeping emperor chicks into the sea before they're ready or forcing thousands of Adélies, which will only breed near sea ice, to abandon old nesting sites. (On the other hand, retreating glaciers on land have made new territory available for Adélies on which to form colonies.)

Even species that breed outside Antarctica, like the king penguin, or those that prefer ice-free waters, like the chinstrap, are yoked to the sea ice and its new dynamism. The Antarctic food web is built on a crustacean called krill, a shrimplike creature about two inches long. Its life cycle is tied to sea ice and the algae that grows beneath it, which krill scrape off to eat with their tiny whirring forelegs. At their highest densities, krill might gather by the millions, by the billions, with as many as ten thousand individuals per square yard. The reddish smears they form extend for miles. They can be seen from space. Whales gorge on them. Penguins do, too.

Krill abundance varies naturally, but shrinking sea ice cover, on top of a growing fishery that hauls in more than one hundred thousand tons of krill each year (mostly to make aquaculture feed), have led to declines of up to 80 percent in some parts of the southern oceans. Thus, even though they were predicted to benefit from less sea ice, populations of chinstrap penguins decline as they have a harder time finding food. King penguins could potentially suffer as well, since they feed on small forage fish that in turn feed on krill.

For penguins in the desert, the effects of climate change come in different ways. One way is rainstorms—their timing, their frequency, their strength. Since 1982, Dee has watched the weather here, keeping track of daily temperatures (high, low, current), rainfall (usually zero, but, as we have just seen, not always), cloud cover, wind speed and direction. Storms would hit from time to time, sometimes with dramatic results; one especially bad one in 1999 had killed sixty-seven study chicks. That big storms were a fact of life (or death) for seabirds was known, but Dee wondered if a broader pattern lurked under all the anecdotes. Were there more storms at Punta Tombo now than when she began? As can happen when her curiosity is stirred in such a way, she decided to find out. ("Dee does that a lot," Ginger told me once. "Ideas will be floating around, and then she'll come in and say, 'We have to work on this right now.'")

In Seattle, Dee and Ginger went back through all the project's weather data. Ginger also found records from the airport in Trelew, which went back decades. In the twenty-eight years they looked at, storms killed chicks in thirteen of them; or, put another way, storms killed eight percent on average of whatever chicks were alive when they hit. In two seasons, storms killed more chicks than anything else: 43 percent of all chicks in one year; in the other, 50 percent. (Lest I get too depressed about so many chicks dying in 11.1 millimeters of water, one chick died during a storm when it rained just 1.2 millimeters.)

There was also the nature of storms. Big ones that struck early in the season, when chicks were small, tended to be deadlier, and Dee and Ginger found the frequency of such storms had increased since the 1960s. The Patagonian steppe was getting more of its rain from these bigger, fiercer storms, and climate models predict the trend will only continue, with increases in extreme rainfall events of 40 to 70 percent within this century alone. The penguins, not having had to adapt to such events, will likely suffer as a result.

It was one of the first studies to show a direct effect of climate change on a seabird. Here, Dee said, is what climate change looks like to the penguins of Punta Tombo: water, more water, too much water, and too soon, before these waterbirds are ready for it.

⤸

The sky has cleared by late afternoon. What few clouds remain look like big, fluffy piles of cotton, but we know the truth, having spent a long day in the Factura and Max-Vista areas, accounting for both the living and the drenched dead. As we visit the rest of the areas over the coming ten-day circuit, we will find more chicks that likely died during

The colony before a storm rolls through.

the storm, although their advanced decay will make it hard to tell what killed them.

Now, sitting on the crest of the hill, and looking over the colony as it dries out, a couplet by Ezra Pound comes to mind:

> The wind is part of the process
> The rain is part of the process

The lines are from one of his longer cantos. He scribbled them on a piece of toilet paper in 1945 while the American military was holding him in detention near Pisa, Italy, for pro-Fascist radio broadcasts he had made during World War II. Imprisoned in a wire cage, exposed to the elements, and slowly losing his mind, he used the wind and rain as a refrain. They were a touchstone, a reassuring sign of the natural order of the world. But not here. This has been a wind and rain without instructions.

We are about to leave when Emily asks, "What's that over there?"

I look where she points. Some forty or fifty yards away, a large white thing is lying on the campo. It might be an enormous kelp gull, but when we arrive, we see it is something much more marvelous: a black-browed albatross. Its eyes are open in the sightless way of the dead, and the black brow of its name gives it a stern, dignified expression. Its white body is bright and spotless. Its dark wings are tucked against its sides. When I lift the body, the wings gracefully unfurl, tumbling to the ground.

The black-browed albatross is the most common albatross in the world. More than a million live throughout the southern hemisphere, although their numbers have been dropping. (They flock to longline fishing boats and trawlers to scavenge, and they pay dearly for it. Worldwide, they are thought to be the seabird most frequently killed from entanglements in fishing gear.) Unlike the penguins, they have no colonies on the Argentine mainland, so while we know they are out there, they have yet to come close enough to the land for us to see them.

We gather around the albatross. We have no idea how it died, but I can't shake the belief that the storm snatched up this bird known for its casual mastery of wind and cast it to the ground simply to flaunt its own power.

"It's beautiful," Briana says.

"What should we do?" El asks. No one is sure. This is a contingency even the almighty protocol has failed to anticipate. We respond the only way we know how: Briana takes out her GPS device and records the albatross's coordinates in her notebook; I pull the scale and small ruler from my pocket. The ruler is perfectly adequate for a penguin, whose flippers are about six inches long; less so for the full length of an albatross's outstretched wings. But it's what I have.

We measure and weigh the giant bird as best we can, and then fold its beautiful wings and lay its body back under a bush. I tear off a strip of bright pink flag and tie it to a branch. The albatross will probably be gone tomorrow, dragged away by the *colpeo* fox or some other scavenger of sufficient ambition and hunger and strength. We gaze at the thing a moment longer, and then crunch to the tourist trail and start back to the house, leaving the long pink flag to flap behind us in the breeze.

7

Hunger

A couple of days after the storm, Turbo leaves to forage. He comes, he goes; it is his way and we are used to it, but I miss him when he's gone. I miss how he runs out to greet us every morning and night when we pass his bush on our way to Cañada. (We go twice a day now that there are chicks.) I miss his distinctive patter, the way he cocks his head and thrusts out his chest, *cuck-cuck-cuck*ing. In my mind, he is singing out, "Helloooooo, friends!"

The lengths of his absences have varied. Sometimes he is away for a few days, sometimes a week or more. In the middle of October, he left for an entire month. What an anxious time that was! I started to fret. "He'll come back," Dee said, and he did.

This time he returns after about a week. "Turbo!" I exclaim, as he scoots out from under the barbed wire and toddles over. It is a little after nine o'clock in the evening. He does his circle dance around my shins while I root around my vest for the scale. Dee has asked us to weigh him after he returns from his forays, but I'm in no rush and squat down next to him. He stops and lets me scratch his neck. Since he was just at sea, his plumage is clean and sleek. I work my fingers into his thick feathers, bury them down to his rough skin. He closes his eyes, and we sit like this for a moment, enjoying the dying light of the sun, while out on the campo, the rest of the penguins carry on.

Oh, Turbo. We've had a rough week.

〜

Different penguins use different strategies to ensure that the world will always have penguins. Two species, the emperor and the king, lay one egg per year. They devote all their energy to it. If the egg is lost, or the chick starves or is eaten, so be it. They will try again next year.

The rest of the penguins lay two eggs, but they don't all have the same goals for them. Among the *Eudyptes* penguins—the crested penguins, that is, such as the rockhoppers or the royal or the Snares—the first egg can be up to 57 percent smaller than the second egg. Next to the second egg, it looks like a Ping-Pong ball. The disparity between the two eggs is the most pronounced of any bird, and the crested penguins' investment in the first appears cursory at best. With one species, the northern rockhopper, both eggs hatch in only 6 percent of nests, and if a chick hatches from the smaller egg at all, it usually dies within a few days. Why the crested penguins do what they do continues to puzzle biologists. Some think the first egg might act as a spare, should the second be lost. Or it serves to prime breeding behaviors: when the female lays the second egg, both she and her mate are in the proper mood to care for it, thanks to the first. It could be due to the crested penguins' long migrations: since the females begin ovulating while still at sea, they may not have the resources to invest in proper egg development. Whatever the case, I wonder whether we have caught crested penguins in the midst of some great evolutionary transition, as they move from laying two eggs to one. But I will not be alive in ten million years to find out. (Neither will they, if current population trends hold.)

Magellanic penguins, like the others in their genus, lay two eggs, as we well know. Unlike the *Eudyptes* penguins, their eggs are about the same size, so the parents have similar hopes for both. Again, we see how finely the penguins' lives are honed to achieve ambitious if elusive ends. A Magellanic female might have laid her eggs four days apart, but she doesn't fully incubate the first until after she has laid the second. The chick from the second egg thus hatches only two days after the first; in absolute terms, it has developed more quickly. The first chick will still have a head start—a two-day lead means a lot if you weigh less than three ounces—but it isn't as significant as it could have been.

∽

When eggs in a clutch hatch on different days, as the Magellanic penguin's do, this is known as hatching asynchrony. The chick that hatched first will be bigger than the second, which is called a size hierarchy. In the 1940s, David Lack, a British ornithologist, wondered what the benefits of such size hierarchies might be. Many birds, like robins or sparrows, incubate their clutches so that all their chicks hatch at the same time and so are the same size. Others, like penguins, don't. Why not?

Lack reasoned that, independent of predation or other unforeseen events (such as storms), birds can raise only as many chicks as they can feed. For many birds, and penguins in particular, food is always an unpredictable resource. When they lay their eggs, they don't know how much food there will be. In some years, there might be a lot, in other years much less. Magellanic penguins therefore live in what is called a food-limited system. In such a system, the surest way for birds to raise as many chicks as possible is to use chicks' asynchronous hatching to establish a size hierarchy in the brood. Per Lack's model, the parents feed the larger, healthier chick at the expense of the smaller chick. In the years when food is plentiful, the smaller chick might get enough to survive, and both chicks leave the nest, or fledge. In years when food is scarce, the smaller chick starves. This is called brood reduction.

Taken as theory, brood reduction is a sensible strategy: feed the bigger chick, give whatever food is left, if any, to the smaller chick, and a pair of penguins has a decent chance of raising at least one chick. This is better than trying to feed both chicks equally, watching them both die because neither gets enough food, and raising zero chicks. Studies have generally borne out the brood reduction hypothesis in its broader strokes, but caveats and provisos add filigree. Brood reduction helps explain how Magellanic penguins balance a desire to send two chicks out into the uncertain world against the likelihood that only one will survive to fledge.

Brood reduction also makes plain how much of life is disposable and subject to chance. When Dee was here, she sometimes likened chicks to lottery tickets. "Having two chicks is like having two tickets," she would say. "You have a better chance of winning than if you have just one." That may be true, but it doesn't change the fact that your chances were never that great to start with, and at Punta Tombo, this conversion from the theoretical to the actual manifests in an especially hard light. I knew about brood reduction before I came here, but I had never seen it. As a felt reality, I was learning, it is awful.

⌐⌐

Once Turbo and I are sated from our bout of mutual preening, I loop the strap under his flippers and attach it to the spring scale. He observes all of this coolly. "Sorry, buddy, science must have its meat," I say, and up he goes. He sways placidly and gazes around, taking in the colony from three feet in the air instead of his usual height of about two, before he gets antsy and starts to wriggle and kick his feet. Poor fellow. He doesn't like being weighed, but such are his concessions for our company.

As he tends to be, he is good and healthy, and heavy enough that my arm starts to ache and tremble from holding him up. At last the scale centers. Turbo may be several days removed from his last meal, but he still weighs 4.95 kilograms, or just shy of eleven pounds. Whatever his failings at conspecific recognition, he is a penguin of exceptional quality when it comes to finding food—better than a lot of the penguins that have mates or chicks. (It also helps no doubt that he hunts only for himself.) But he has had enough. He leans forward, grabs my sleeve with his bill, and gives it a firm, remonstrative twist.

I put him down and free him from the strap. He patters away. "Sorry!" I call after him. He doesn't look back. In treating him like any run-of-the-mill penguin, perhaps I have offended him. He doesn't feel like being data today.

⌐⌐

That the storm scythed through so many chicks so quickly may have been a shock to us, but chicks had been dying for a while. Death for them was a garden with many forking paths: they might be snatched up by kelp gulls, torn to pieces by *peludos*, stomped on by their parents, trampled more generally, impaled on branches, crushed during fights, neglected, baked in the sun, drowned in puddles, or smothered when their burrows caved in. The plurality, though, simply starved to death. Dee has found that about 40 percent of chicks starve each year on average, and sometimes as many as 86 percent. Most starve when they are young, mortality sharply peaking when they are between five and nine days of age. Dee suspects that, more than the outright favoritism brood reduction would predict, the chicks are victims of bad timing. Once they hatch, they have about a week of nutritional reserves left from their yolk. They need a meal before those reserves run out. Often, the parent out foraging just doesn't return in time, and the attending parent has no food to offer. The days pass and the chick waits, and waits, and weakens, and waits, and weakens, and waits, and then dies. Or, even more cruelly, sometimes a parent *does* return after a few days, but the chick dies anyway. In the absence of food, its digestive system hasn't fully developed, and so the meal sits in its stomach, worthless.

I suppose I shouldn't have said *simply starved*. Starvation is never simple. Of all the ways to die, it is the most drawn out. We soon can tell which chicks are not long for this earth. They are the ones guarded by the same parent day after day. ("When is your mate coming back?" we plead.) The ones that gain almost no weight between our visits, or even lose a few grams. The ones that beg in raspy, thin voices, while their parent can do little but head-wag at us. The ones whose feet become shriveled and translucent, the ones whose thin bones we can feel through the sagging skin of their chests, the ones who lie quietly in the dust, breathing shallowly, waiting for the end. These are the most piteous to see, and even harder to measure in the knowledge that the next time we visit, we will find their rotting bodies, if something hasn't already made off with them.

More troublesome are our own contributions to the chicks' already considerable distresses. Daily we make our rounds through the areas, plucking out those chicks that have survived the week, the night. We measure and weigh them. With Sharpies, we color the older chick blue and the younger one green, so we will know who is who next time. When

Dead chicks pulled from several study nests. The shriveled feet and flippers indicate starvation was the likely cause of death. We remove the bodies so they don't attract scavengers, which might in turn attack either surviving chicks, or living chicks nearby.

their feet are big enough, we insert a small metal tag through the webbing on their left foot, assured it is no more painful than piercing an ear. Those chicks that live grow quickly, both in size and subjectivity. Their eyes open, and they start to see us for what we are. In response, they wail and shit on our pants.

The work starts to seem to me an unpleasant mix of death and life, of aid rendered obliquely but discomfort directly dealt. It is the paradox of Punta Tombo. We are in a massive breeding colony, one of the largest in the world. Life should be streaming out of here in waves, so much life, but instead there is mostly suffering and a disinterested biological cruelty that grasps at everything with its bony fingers. A few days ago, El and I were walking back to the house when we came across a female tinamou leading her passel of chicks. "Look! Look!" El said, brightening, for it had been a hard day. The tinamous saw us at about the same time. The female gave a little *squawk!* and she and her chicks scattered. There were eight or ten of them. They were small and fluffy and very fast. They skimmed over the ground on blurring toothpick legs. Some sought shelter in bushes, others crouched behind rocks. One scurried into a penguin burrow, where a male was dozing with his chicks. The little tinamou went to hunker down next to him. When the penguin saw the tinamou, he leaned over and nonchalantly snapped its neck, killing it.

El and I stopped and stood as if turned to stone. Our delight evaporated. "Oh, my god," El whispered. "What did we do?"

That was the first time she cried.

⤳

The second time she cried was over a chick named Verbiage. Verbiage had hatched in late November, the first chick in a Sea-Tip nest, 301V, under a small *ojo de vibora* bush, to an old female, band number 48196, and her mate, t6672. (The "t" stands for *toe tag*, like the ones we put through the chicks' feet.) They had been a pair for three years.

Two days after Verbiage hatched, El and I came to the nest in the evening and saw t6672 brooding his second chick, which must have hatched the night before. We removed and measured this new chick, and as I returned it, we tried to think of a name for it, per our usual practice. Good V names were hard to come by, and while we mulled, I extracted Verbiage, too, so we could weigh the chick, as the protocol instructs. t6672 was becoming more agitated. He tore at the duct tape wrapping on the eggcup as I coaxed Verbiage in, shredding it with ease. He seemed to me especially fierce; or, as we would euphemistically say, protective.

I weighed Verbiage. By now, t6672 was huffing and snorting. When it was time to return Verbiage, I distracted the male so El could slide the eggcup under him, which she did, placing Verbiage right between his feet. It was a deft maneuver, and for a moment we thought a successful one, but then t6672 saw Verbiage squirming. He grabbed his chick and beat it against a rock, thrashed it back and forth, and spat it out. Verbiage tumbled into the dirt, shuddered once, and died. The chick's life, a breath, was gone.

His foe vanquished, t6672 turned to El and me and brayed savagely. El gasped and sagged to the ground and started crying. I reached into the nest with the eggcup to retrieve Verbiage, but before I could get the body, t6672 neatly eviscerated it. Something yellow bubbled out of Verbiage's belly. I was furious. I shoved t6672 roughly aside and grabbed Verbiage with my bare hand. "You stupid fucker!" I yelled, but I knew my rage was gratuitous, directed as it was at something with limited powers of choice.

I sat with Verbiage's guts oozing over my hands. El was still crying. "What should we do?" I asked desperately. We had caused this. We had to do something, anything, but I had no idea what. El took a minute to collect herself, and sniffled. "Let's put her on the berm," she said.

The berm was only a few yards from the nest, with a line of small bushes growing along its crest. They looked like bonsais, stunted and twisted from the wind. None was more than a foot tall. We found one with space under its canopy and set Verbiage there, arranging the chick so it faced the sea. The wind blew steadily but gently. The sun was setting behind us, painting the colony a rich rose. Below, hundreds of penguins went about their business on the beach, as the waves advanced and withdrew, advanced and withdrew.

We sat with the body for a while, and then El looked at the nest page where I had written what would become the official account of this little tragedy: ♂ *pecked at #1C when returned to nest, then thrashed it to death. COD = parent.* (I'm not sure now the last part is entirely accurate.) She gazed at the sea before reaching some conclusion. Next to the measurements of the surviving chick, she wrote in her neat hand, "Violet." Then the pencil fell from her fingers and she started sobbing. She sobbed and sobbed, and I just sat there like an idiot. I didn't even have the husbandly wherewithal to put my hand on her shoulder.

We got up to walk back to the house. As we crunched along the tourist trail, I heard El quietly say, "I hate science." Her antipathy had been building for some time. She wasn't trained as a scientist and hasn't been taught, as I have, to accept the necessity of our work without question, to say nothing of its nobility. She evaluated everything we did according to her own criteria, guided by her radical empathy. I had sensed her reservation the first time we made a penguin bray in terror, had felt her skepticism when Ginger assured us that our stern handling of the penguins was fine because, as she so often said, "It's for your own safety, and anyway you cannot hurt these birds." I knew El wouldn't necessarily privilege her own safety over that of the penguins, but she had pushed her objections aside then. No longer. She went on, still calm, partly to me, mostly to herself. "I hate that so many of the chicks' lives are so brief. I hate that during their time they know nothing but pain, and I hate the things we do that add to their pain."

I knew how Dee would have responded: Things happen. We had returned the chick perfectly well, and there was no way we could have known the male would react the way he did. He probably would have killed his chick before long even if we had never been there. Dee would have told us to remember that we're here to help the penguins, and however much they're afraid of us, in the end we're doing them more good than harm. Much more good. It isn't even close: the results of this project will have ramifications that extend far beyond the shores of Argentina, and that, in the face of inevitable and unfortunate and even heartrending setbacks, is why we have to keep going. But those rationales seemed an insufficient answer for Verbiage's death, so I said nothing. There is a difference between a death in matter and a death in the spirit, and I wasn't competent to address such things.

We walked the rest of the way to the house in silence, and El went inside while I put away our gear. I could hear Briana and Emily tending to her—at least someone knew how to respond—but sitting on the concrete skirt next to the truck, I felt wretched and useless. I had brought El here, had promised her it would be fun, an adventure. The work could be difficult, true, but we were trying to see Patagonia as something other than hard and dry; or, god forbid, bleak, that awful cliché for this region. Yet here she was, more miserable than I had ever seen her.

I thought of *Attending Marvels*, George Gaylord Simpson's account of his time digging for fossils near here in the 1920s. Early in the book, there is a passage when everything was going wrong, and he gave in for a moment to despair. "It was bleak," he wrote. "'Bleak, *a.*,' says the dictionary. 'Exposed to wind and weather. Syn.: bare, barren, bitter, blank, cheerless, chill, cold, cutting, desolate, dreary, exposed, hostile, raw, stormy, unfriendly, unsheltered, waste, wild, windy.' The dictionary has taken the words right out of my mouth. Patagonia is all that, and I may as well start repeating it now: Patagonia is bleak."

Yes, Patagonia is bleak. There, I said it, too.

The landscape of The Point.

That night, after El had fallen into a deep sleep, I slipped out of the trailer to go for a walk on the tourist trail. I had started doing this from time to time as the season took its darker turn, roaming about in the wee hours. Ostensibly, my purpose was to count the fishing boats—there are few activities from which we can't somehow glean data—but in truth I just wanted to be in the colony in the dimmest light. It could be calming to walk among hundreds of thousands of penguins without seeing them.

Chickless, girded in his armor of the I, Turbo ran out to meet me as I skulked past his bush. After a hasty hello, I dodged around him and jogged away. He followed for a bit before giving up and turning around. I watched him go. Happy as I usually would have been to see him, I was starting to question the nature of our relationship. I worried I was using him, that in depending on him to sooth my unsettled feelings while dismissing the attachment as false, I was trying to absolve myself from having to look at all the penguins and their individual plights with an equivalent compassion. What ought to have been a broader and more diffuse empathy was focused instead on a single batty penguin.

I walked to the crest of the hill and headed deeper into the colony. Penguins were braying all around me, but mostly what I heard was the steady clamor of begging chicks. With their stereophonic *peep*s came a battering of questions. Do I not care enough about them? Why don't I care as deeply as El does? Is it good to try to hold despair at bay? What is the best way to balance empathy with ecology more broadly? Must one do obeisance to the other? More broadly still: Is all of life nothing more than algorithm—nasty, brutish, and short?

"Lives that fit the natural scheme take continual loss," the naturalist John Hay has written. When I read that years ago, I thought I understood what he meant, but I never felt his full meaning until the penguins of Punta Tombo showed me as often as they could, mashing my face against their wildness in all its naked mechanisms. Theirs was proving a devastating lesson.

I walked along and listened to my feet on the gravel, took in the breadth of the sky, its unfamiliar constellations. The moon was nearly full, and so bright that even from the tourist trail I could see the spume veiling off the waves as they rolled onto the beach. But was it sea mist? I peered at it. No, something else. With every wave there was a flare of light, a blue flare. Up and down the coast, I now saw, the breakers were glowing an eerie blue, wave reiterating upon wave. When they crashed into the beach, the blue exploded all over.

I ran down the tourist trail to the beach, sprinting as fast as I could. Penguins scrambled out of my way. I reached the end of the trail and stood heaving. Closer to the sea, I understood better what I was seeing. The blue water-light was caused by a bloom of bioluminescent phytoplankton. When disturbed in some way, the plankton emit a tiny pulse of phosphorescence, so that waters full of them seem to glow. How long the bloom had been here I had no idea, but since we only looked at the ocean during the day, we had never seen them.

I found a good-sized rock and hurled it as far as I could, thrilling to the blue eruption of its *sploosh!* A handful of gravel sparkled like a firework. Against this, the waves pounded out their steady blue rhythms.

Wonderful.

Then I saw a bluish glow about two feet long, a streamlined form, moving swiftly through the water toward the beach. I stared at it until

I realized what it was: a penguin swimming in from the sea. The form wavered and blurred, and I could see brief flares of blue when the animal pumped its flippers. Faint vortices swirled in its wake before fading in the black water. Another penguin streaked in a few seconds later, this one much quicker, and then another. A steady flow of them was returning from foraging, swimming in one by one. Most swam directly, but some meandered, noodling around a few dozen yards from shore before suddenly turning and zipping in. In shadowed clarity, they emerged from the water and started the slow, metronomic trudge up the beach to their chicks, earthbound once more.

I had never seen anything so strange and beautiful.

Penguins were leaving the colony, too. Unlike those returning, these often went in little groups of two or three, or sometimes four. They swam from the land swiftly and directly, shooting off like comets. Just when I feared I couldn't possibly stand any more amazement, a great smear of light moved from the beach into the water. It stopped for a moment, the glow hovering on the swell, and then it burst into shards of light as seven penguins swam off on their separate ways.

I have read that three things, miracle, mystery, and authority, can be used to quell certain types of spiritual rebellion. Here I was in the thrall of miracle and mystery. (Authority, in the form of Dee, presided from Seattle.) Something inside me that had been under tremendous strain gave way. I stood on the rocks and laughed, yelled, howled, careering between states of dizzying ecstasy until I wasn't sure what I was doing other than making loud noises. I was glad no one else was there.

I must have watched the glowing penguins for more than an hour, but the time passed without my knowing it—an hour collapsed into a second, which itself was stretched into an hour, thirty-six hundred times. I watched in thoughtless rapture, and when I did eventually think, the first thought that came to me was that the ghostly blue shimmers leaving the colony were the souls of dead chicks swimming away to whatever penguin afterlife there might be, and the bright birds arriving were the souls of chicks about to hatch, that we might measure tomorrow. I watched all these souls fly into and out of the colony. I knew I shouldn't think such things, that it was symptomatic of maudlin tendencies I am always trying to suppress, but Punta Tombo was turning me into a simple soul. Show me something dark that can blaze with its own light, and I will call it magic.

8

Watching Seabirds at Sea

Dee is back. She will be here for about four weeks, until early January. Arriving shortly after the storm, and on the heels of all the dead chicks, she finds us more somber now than when she left us, heavier in our movements.

"How are you all doing?" she asks one evening. She and I are "counting left flips," as she calls it, which is to say we are strolling through the campo and *click*ing penguins, looking for banded birds. (Since bands are on the left flipper, we *click* only those penguins whose left flippers we can see.) If we find one we haven't seen yet this year, we note its location, but we are in Doughnut, and all the penguins so far are known to us. Really, counting left flips is a way for Dee to enjoy a walk at the end of the day and call it work.

Click click. Click. Click click click.

"We're alright," I say after a moment. I'm not sure how much she wants to know, how much I feel like telling her. "It was hard for a while." *Click.*

"It can be," Dee says. *Click.* "How's El holding up?" *Click click.*

"She's okay. Some good days, some not-so-good days."

"I know it's tough when the chicks die," Dee says, "but it could be a lot worse." *Click.* She's right, of course. Back in Seattle, a printout on a wall tracks the colony's reproductive success over the years. (Reproductive success is the number of chicks that fledge per nest that had eggs.) The average is just above 0.5, or half a chick per nest. This means that in a standard year, three-quarters of all eggs and chicks will be lost. In good years, reproductive success might climb as high as 0.95. In bad years, it will sink closer to 0.3. In soul-destroying years, like 1999 and 2000, when storm after storm slammed into Punta Tombo and there was hardly any food, almost nothing survives. So yes, it could be a lot worse.

Click. Click. "I've been impressed with El," Dee says. *Click.* "She's a good worker. She's really taken to this."

"Yes," I say. For Dee, there is no higher praise than "good worker." I also know she isn't impressed by too many people, and I count myself among those she considers wanting in some way. *Click click.*

"You really lucked out," she says. *Click click.* She chuckles. "Good mate choice."

"Thanks." *Click click click.*

Around us, penguins bray, waddle, preen, nap, brood their chicks, do nothing. This could be any evening from the first three months of the season, except now the air is suffused with the sweet stink of death and decay. Dee eyes a penguin that might have a band, but what looks like a steely glint is only a trick of light. *Click.* "We're too insulated from death," she says suddenly. "We forget how much a part of life it is—things are born, things die, that's the way it's always been. Here you can't get sentimental. The penguins won't let you." *Click click.*

⟿

I don't know about sentiment, but I'm learning a lot about continual loss. El and I have been making regular trips to the berm. A few days ago, we brought up a chick she had especially loved from Cañada, which added half an hour to our day since Cañada is pretty far inland. (Usually, we buried those chicks outside their nests and decorated their graves with colorful pebbles.) We brought up a chick whose parent returned with food right after it starved. We brought up a chick that died for no apparent reason other than that its little spirit was unequal to the demands of living. It is a dreary business. We are running out of stunted trees.

That said, if the chicks are still dying, at least they are no longer dying at such a dismal rate. The pulse of early mortality has passed, and the more competent parents have settled into a cycle of frenetic provisioning.

I think of these penguins as the elite of Punta Tombo. Their diligence cheers me up, as do their chubby offspring. We often find these chicks lolling in a stupor of surfeit, their distended bellies hard as rocks. "Tight as a tick!" Dee exclaims whenever she handles an especially fat chick. "You're just as tight as a tick!" Some chicks are so round that when we finish measuring them, we simply roll them back into their nest before the eyes of their astonished parents. Dee calls this "chick bowling."

So the chicks get bigger and bigger and bigger. Some of them triple their weight every ten days. Their flippers are longer, their feet stronger. When they were young, at least one parent was always with them, but by mid-December, most are large and mobile enough to fend for themselves. Their parents start to leave them alone at the nest, as both mother and father spend as much time as possible hunting for food. This marks the period Dee calls Late Chick. (The double entendre just occurs to me.)

The penguins come, the penguins go. We who are left behind note their movements as best we can, but our view is necessarily limited.

A chick trying to hide while the author measures it. As chicks grow, they become considerably less helpless.

When I drop by a nest and find the chicks home alone, I know only that their parents are gone. Of the essential information encrypted in their absence—how far they are going, how hard they are having to work to find anchovy and hake for their young—I know nothing. These questions are one of the reasons Dee and I are out walking and *click*ing this evening: we are looking for a penguin to carry a satellite tag so we can follow it to sea.

~

Penguins have at best an ambivalent relationship with the land. They are never more vulnerable than when they are on it, never more awkward than when they stumble over it—but try to lay an egg on the water. So penguins, and all seabirds, are compelled to come ashore from time to time. Land is where bird and scientist most often meet, but while the colonies may be logistically convenient, we, the scientists, have always wanted to go where a seabird's natural gifts are most gracefully displayed.

One of the first scientists to try this was my bard of the campo, Robert Cushman Murphy. In July 1912, at the age of twenty-five and married not even a year, he joined the crew of a whaleship called the *Daisy*. He would sail aboard her for ten months as she searched the Atlantic for cetacean quarry. Setting off from the Dominican Republic, she ranged far and wide, first heading north, then east to the Cape Verde islands, before turning south and west and meandering off the South American coast until she reached South Georgia Island, just above the Antarctic circle. She stayed at South Georgia for several weeks, until it was time to go back north.

Being at sea may have meant "leaving the world," as Murphy would say, but that was just what he wanted, to immerse himself in the marine lives of birds. Since there was no recognized title for scientist, he was listed on the ship's manifest as an assistant navigator. He was the lowest-ranking member of the crew, entitled to just a 1/200th share of the revenue, or half a penny on the dollar. A tall man, he spent hours on the deck scanning with his binoculars, noting what species were where and making guesses as to why. He collected (that is, shot) hundreds of individuals from dozens of species. He dissected and cleaned their bodies, made copious measurements of their various parts, filled notebook after notebook with data and sketches. He salted and packed the skins away so they wouldn't rot on the trip home. When circumstances and the captain would permit, he rowed out in the *Daisy*'s dinghy. Just out of sight of the ship, he rode the waves and sat among the birds as they floated around his dinghy, or wove through the air above him.

The *Daisy* returned to Barbados in May of 1913, and Murphy went home to New York. He took with him his earnings of a few dollars, and more than five hundred skins from fifty-five species. He would go on to gain renown as an ornithologist, becoming the curator of oceanic birds at the American Museum of Natural History. A petrel is named after him (Murphy's petrel, *Pterodroma ultima*), and also a species of feather louse that infests albatrosses (*Eurymetopus murphyi*), as well as a spider, a plant, a fish, a mountain in Antarctica, and a junior high school in Stony Brook, New York.

In 1936, twenty-three years after his voyage on the *Daisy*, Murphy published *Oceanic Birds of South America*. As the first major treatise on seabird ecology, it was his answer to the prevailing assumption that seabirds flew (or, if a penguin, swam) hither and yon with little regard for local details. The two heavy volumes contain, in addition to hundreds of detailed life histories, chapters on hydrology, biogeography, meteorology, and ocean currents. Throughout, Murphy argues that the ocean was a habitat as highly varied as any on land. It "abounds in invisible walls and hedges," with stark differences in temperature and salinity. Seabirds therefore distribute themselves per their needs and tastes, and, he wrote, "are bound as peons to their own specific types of surface water."

Murphy's work was driven in part by the fear that if he didn't document seabirds' lives, the chance would soon be lost. He saw the ocean as the next frontier, the "true agent in the fate of nations," as one historian put it. He had watched with a mix of horror and awe at the efficiency with which even a crew as small as the *Daisy*'s could slaughter its way across the seas, killing almost everything they happened upon. Such despoiling brought to mind another ravaged wilderness. As a younger man, Murphy had read of the fall of the American West, victim to its most precious myth, the myth of infinite resource. He didn't want the oceans to share that end. In a speech he gave to the Garden Club of America near the end of his life, he said, "The idea of the new term 'proper land use' must, of course, extend its meaning to the sea."

$$\backsim$$

Once Dee and I have *click*ed for a while, I ferret out the Doughnut book. In it, I've paper-clipped the nest pages of a few promising candidates for satellite tags, as determined by a set of loose criteria.

1. The satellite penguin's nest must not be so cramped as to endanger the tag's three-inch antenna.

2. Bushes are generally better than burrows in this regard.

3. The closer to the house the nest is, the better, since we will check it twice a day; and up to fourteen penguins will have satellite tags at any one time.

4. The penguin must be a male. Although Dee deployed tags on both males and females in the first years, the sexes behaved similarly, so she decided to remove sex as a variable.

None of these conditions seem terribly hard to fulfill, but I have yet to learn the gestalt of the ideal penguin. One bush nest has too many branches hanging over the entrance. "They could snag the antenna, and there goes five thousand bucks," Dee says. The opening to another burrow isn't too squat nor snug, but its walls slope in a complicated way. "If the penguin backs in, the antenna could get mashed," Dee says. Another bush nest isn't quite right for some elusive reason. "I don't know, this one just isn't good," Dee says.

Dismissed paper clip after dismissed paper clip falls back into my pocket. "What else have you got for me?" Dee asks.

I take her to Doughnut 223L, a bush that is home to a not-known-age male (57577) we banded a couple of months ago. As a nest, his isn't the greatest. The entrance is large, and a scraggly thatch of branches covers the nest cup. All a flying kelp gull has to do is look down to see the contents; and without shade, the chicks are liable to bake on the hottest days.

"This one looks good," Dee says. She tugs at the branches, breaks a couple of the smaller ones off. Her approval strikes me as another of Punta Tombo's little incongruities. Generally, the better the nest, which we define as having the greatest cover, the more chicks the pair occupying that nest will raise. To carry that logic further, better penguins (bigger, stronger, faster, whatever) should have better nests. Yet Dee doesn't want too great a nest for a satellite penguin, and therefore not too great a penguin. She seeks a perfect mediocrity, and here is 57577, so middling as to be extraordinary. He watches Dee take the measure of his life choices. "Yeah, this one will work" she says at last. "We'll come back tomorrow and put a tag on him."

57577 head-wags at us as we leave.

After Murphy, research on seabirds at sea would languish for the next several decades. This was due more to a lack of funds than a lack of interest. Most government research organizations won't underwrite a ten-month trip on a sailing ship as an assistant navigator, and biologists had to be creative in their search for support. During the 1960s, a few approached the military. They wanted to try to track seabirds remotely and argued their work could help with the development of missile-guidance systems. The US Army gave some money, but for a different reason: officials wanted to know whether seabirds might be vectors for disease.

It wasn't until the 1970s, when countries began to establish Exclusive Economic Zones, that research of birds at sea became more widespread. Exclusive Economic Zones, or EEZs, are the claims a country makes on all the marine resources that surround it, which is to say primarily the oil and fish. They extend two hundred nautical miles off a country's coastline. Those governments with more robust environmental regulations were obliged to learn what animals were also using the EEZs. Scientists were thus sent out to survey them, ushering in, as one researcher would later write, "the golden age of at-sea studies of seabirds."

Most surveys were done from ships, until the early 1990s, when biologists started to use small planes. Both ship and plane could go where the seabirds were, but as methods they still depended on the chanced glimpse. Ship or plane would follow a set transect, going out and back, out and back, and surveyors would note whatever seabirds they came across. They had no idea where the birds had come from, how long they would stay, or where they were going, but on such-and-such a day at such-and-such a time, here they were, and that was better than nothing.

⌐

By the early 1990s, Dee had been coming to Argentina for ten years. Her terrestrial protocol was well established, and she was ready to devote more of her energy to the sea and its meanings. Her first question was where penguins went when they foraged during the breeding season. While at Punta Tombo, penguins are what is known as a central-place forager: they leave from and return to a single place, their nest. When a central-place forager has dependent young, they are constrained in the maximum distance they can travel from that place to find food. Swim too far, and they risk not having enough food left undigested for their chicks when they come back, or worse, their chicks starving to death while they are gone. Physiology and time combine to enforce a spatial limit. It would be a useful limit to know if one wanted to advocate for a marine protected area, which Dee did.

To figure out what the limit might be, Dee first went to the Oregon Zoo in Portland. The zoo had a small flock of Humboldt penguins, a close relative of the Magellanic. Dee fed the penguins tablets and monitored the tablets' progress through their guts by X-ray. From this she learned a penguin's stomach takes about eight hours to empty. She reasoned a Magellanic penguin would need half-a-stomach's worth of food to satisfy their chicks, and she knew adults can swim almost four-and-a-half miles per hour. Therefore, she concluded, penguins should be able to forage as far as thirty-six miles from their nest, give or take.

The next time she was at Punta Tombo, she set out to search for all those penguins she expected to be within a few miles of the colony. She got in touch with the Argentine Navy, and a pilot was found who was willing to fly one of their Lockheed Electras low over the open ocean at just above stall speed. Dee's students were eager to help, so to approximate what she thought trying to see penguins from the air would entail, Dee threw black beans on a dark tabletop and had everyone count as quickly as they could. But in the face of the unknown, there is only so much one can do to prepare. On the day of the flight, it was incredibly windy, and the Electra bounced around like mad. The students were supposed to call out whenever they saw a penguin, but they were so nauseous they could barely speak. It didn't matter. Even though the pilot flew almost twenty-five miles from the coast, there were few penguins. On the return to Trelew, Dee watched the pilot tap the gas gauge to cajole the needle to move a little. An empty gesture—the needle remained stubbornly zeroed.

Dee also tried radio telemetry, a type of remote sensing less harrowing than surveying by plane. She attached small transmitters to a few penguins, and then stood on the berm with a radio antenna held aloft, listening for the *beep beep beep* that would mean the birds were close by. She couldn't know where they were exactly, but she would at least have a general idea. Again, she was skunked: the antenna picked up nothing. The penguins were spending most of their time well outside its range. Where had they gone?

Remote sensing still would prove to be the most promising approach. In 1991, Dee joined a colleague in Antarctica who had found a way to track Adélie penguins during the winter by gluing a small satellite transmitter to their backs. The benefit of such a transmitter was that scientists didn't need to be near the penguin, or even know where it lived, to get information. They could simply sit in their offices and wait for the satellites to beam data to their computers.

Dee liked the idea and had a tag designed for Magellanic penguins. It has worked wonderfully, and now she no longer has to wonder where the penguins are. She can know the moment one steps into the waves, and every hour or so thereafter, when the tag sends a signal to a group of satellites whirling far overhead. She can know that penguin's location within a few hundred meters. She can know the temperature of the water that surrounds it. She can know how many times it dives and how deep it goes, as it chases fish through the blue-black depths.

Dee and I are back the next evening. 57577 is still at his nest with both his chicks, I'm happy to see, although they don't look happy to see us.

"Let's get set up," Dee says.

Applying a satellite tag is the closest I will ever come, I hope, to performing a minor medical procedure in the field. We must first prepare the instruments: one tongue depressor and one popsicle stick; a small can of

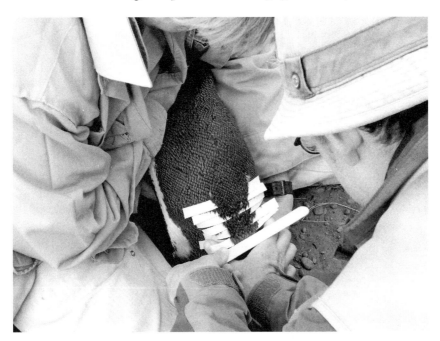

The author affixes a satellite tag to the back of a Magellanic penguin while Dee holds the bird. Several strips of Tesa tape have been wedged under the feathers.

neoprene glue; ten to twelve thin strips of black Tesa tape, each about six inches long; a single strip of duct tape; an empty Qantas bag that once held complimentary toiletries; a small ruler and a paper plate; the epoxy; a foam pillow to sit on; and the tag itself. It has shrunk over the years and so now is only about four inches long and weighs less than 3.5 ounces, but drag is still an issue. A rough equivalent for me of what 57577 is about to experience would be if someone attached a small kite to my back and made me run. Would I notice? Undoubtedly. (They would have had to tackle me in the street to attach it.) Would it unduly affect my comings and goings? Probably not too much.

"Do you want to hold or apply?" Dee asks.

"I'll apply."

"Okay," Dee says. She hauls 57577 out and we weigh him: 4.2 kilograms, or 9.26 pounds. (Adults with chicks are always a little on the skinny side this time of year.) I wrap his bill with the strip of duct tape; he can open his bill enough to breathe, but not so much that he can bite. Next, I slip the Qantas bag over his head and push his bill through a slit

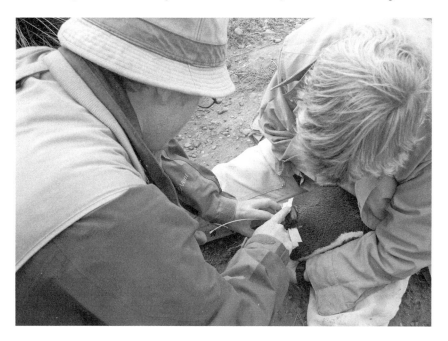

Here, the satellite tag is placed in the middle and then wrapped with the strips. The taped tag is then covered with epoxy to keep water out.

cut in the corner. He struggles, but in an unsure way. He has probably never been in the dark like this.

"All right, give me a second," Dee says. She plops the pillow on a rocky ledge, sits on it, and holds her arms out. "Let's have him." I hand her 57577, and she wedges his hooded head in her left armpit. She uses her body to pin his left flipper and her arm to pin his right, and she cups his tail with her left hand. 57577 jerks and flails and kicks. "There there," she soothes. "There there. It's okay." She keeps talking in a low monotone, emanating a sense of calm I don't normally associate with her; 57577 quiets down, and his breathing slows. Dee nods to me.

This will be the second time we've used satellite tags this season. The first time, in October, I watched Dee apply a couple before she asked me to do one out of the blue. She does this sometimes, just hits you with a request, which is really a command. I never do well in such circumstances, and I didn't that time, bumbling through while Dee held the penguin and barked critiques: "You're putting the tape too far apart! . . . You aren't going to be able to use enough tape! . . . The tag's too low! . . . Now it's tilted! . . . Hurry or the epoxy will set and you'll have wasted the whole batch!"

My face burned and I muttered apologies. "You did fine," El tried to reassure me afterward, but I knew the truth.

Now, I kneel and consider 57577's back, the canvas on which I will work. It is broad and smooth. His body inflates with his breaths. I position the tag a few inches above the oil gland near the base of his tail, and then use the tongue depressor to lift short sections of plumage. I place a strip of Tesa tape and let the feathers fall over the middle third of the strip, with the two ends flapping free like wings. I do this with all the tape strips. I next smear neoprene glue over the patch of feathers, stick the tag there, and wrap it in tape, strip by strip, until almost the entire tag is covered.

"You really mummified it," Dee says.

"You bet," I say. I become aware of an odd *snozz*-ing sound. "What's that?" I ask. Dee grins and nods downward. "Him," she says. Cradled in her arms, 57577 has fallen asleep and is snoring through the Qantas bag. Now she barely holds him. "Listen," she says, clearly tickled. "He sounds like a long-haul trucker."

I had no idea penguins could snore.

Now for the epoxy, which will make the whole arrangement watertight. I mix it and with the popsicle stick spread it over the taped-up tag. ("Like frosting a cake," is how Dee told me the first time, while I slopped it on in heavy gobs.) I dip and spread, dip and spread, and then it is done.

Dee leans over to inspect my work, and I find myself considering the top of her head. In this warm light, at this quiet time, she reminds me of my grandmother, who died a few years ago. Their hair whorled in the same direction, shared the same blend of silver-blonde strands, fell across their heads to form the same soft helmet. I feel an odd swell of affection for this person who has put me through so much, and I mull the influence powerful forces can have on our lives: the sun, the moon, the sea, gods of various stripes, and P. Dee Boersma.

"Looks good," Dee says. "Nice and straight and watertight."

"Great, thanks." I tap the epoxy. "It should be dry soon."

"Good," Dee says. She makes a face and wiggles her knees. "My legs are going numb."

57577 snores on.

⌣

Dee put her first satellite tag on a male penguin early in the season of 1995, during incubation, right before he left the colony after his long fast. When she collected the tag and saw its data a few weeks later, she was stunned. Rather than thirty miles or so, in three weeks the bird had swum more than two hundred and seventy miles, going far out to sea and back. He was no outlier. From 1996 to 2006, penguins would travel on average more than two hundred and fifty miles during incubation, and in extreme cases more than seven hundred miles. That they might swim so far for food is now standard fare in Dee's telling of their natural history, but at the time she had no idea what they were capable of. No one did.

Ever since then, Dee has put on tags—326 as of this year. She deploys at all times during the season—Incubation, Early Chick, Late Chick, Pre-Molt, Migration—and has found a clear pattern. Penguins swim the farthest during the Incubation period, when their mate is incubating the eggs, and they have at least two or three weeks at their disposal. The long, straight lines they draw on the computer screen stretch northeast past the Peninsula Valdés, sometimes all the way to Bahia Blanca, and sometimes farther still, depending on where the fish happen to be that year. But as the season progresses, they must stay closer to the colony, to their chicks. Their absences shorten. During the Early Chick period, when the chicks have smaller stomachs and need to be fed more often, their parents make trips that average less than forty miles each. Their frantic out-and-backs look like little teardrops. In Late Chick, as the chicks' stomachs grow and they can go longer between feedings, both parents can extend their search, swimming an average of almost seventy miles per trip.

The penguins, Dee likes to say, are following the fish. At Punta Tombo, the fish they follow most, as we have seen, are the Argentine anchovy. Anchovy gather in the spring and summer to spawn in fronts of cool seawater. Fronts are natural boundaries in the oceans—Murphy's walls and hedges, if you like. They are characterized by high concentrations of chlorophyll, a proxy for the phytoplankton that form the base of the marine food web. (Phytoplankton attract zooplankton, and zooplankton attract anchovy.) These fronts are what the penguins seek, skilled ocean-ographers that they are. When the fronts and fish are close to the colony, the penguins can stay near. When the fish are far away, they must swim those greater distances to reach them. From their locations, Dee can learn what is happening in the ocean: where the fronts are, how many fish there are relatively. This, she says, is one of the things that makes penguins so effective as ocean sentinels.

The data from satellite tags also helped her learn the nature of their foraging trips. Penguins leave Punta Tombo and head purposefully and swiftly in one direction, but are still open to the possibility of finding food along the way. If they chance across some fish, they might fill their bellies, but then they continue on. It is when they find larger fish schools that their behavior changes markedly. The satellite tracks start to show a jumble of darting and weaving: the tangle of pursuit, capture, consumption. Once they have had their fill, they turn for home. They may have been focused on the way out, but they are resolutely single-minded on the way back, swimming as fast as they can, all day and all night. One penguin tagged during Late Chick swam 107 miles in twenty-four hours, straight from its foraging area back to its nest without stopping. The penguins even modify their physiological processes. To ensure they have fish for their chicks, they will slow their digestion during transit, either by lowering their stomach's acidity or (perhaps) its temperature. By the time they waddle ashore their stomachs are filled with a fishy slurry, which they regurgitate to their chicks.

The greater value of the satellite tags, having revealed the penguins' behavior at sea, may be for conservation. Not only did the tags help Dee identify key penguin foraging areas, but they also showed, and are still showing, how much harder penguins have to work to get food. Dee has found that, over the past decade, the penguins of Punta Tombo are having to swim thirty-five more miles per trip than they did in the past. The costs of this added distance are profound. The farther penguins have to swim for fish, the longer they have to spend eating once they find them, first to replenish the energy they lost and then gathering food for their chicks. The farther penguins have to swim for fish, then, the longer they spend away from their nests; and the longer their mates and chicks have to wait, the poorer their condition becomes. The farther penguins have to swim for fish in a depleted sea, in sum, the fewer chicks they are able to raise.

If the satellite tags paint a grim picture in some ways, however, then they also point to a way to lessen conflicts between human fishers and penguins, who are often in search of the same species of fish. The more we claim for ourselves, the less we leave for penguins and other seabirds. Scientists now argue that, according to their models, we should leave 34.6 percent of forage fish, like anchovy, for seabirds. "One-third for the birds" is how they put it; any less, and there could be catastrophic population declines for many species. This is why, Dee says, the Argentine government should establish a protected area around Punta Tombo and limit fishing there during the breeding season. People don't have to stop fishing altogether, but they can stop for a couple of months to give chicks a better chance to survive, and perhaps this will help slow the colony's decline.

It is the least we can do, she feels. "Life," she has said, "is not likely to get easier for penguins."

⌇

Once the epoxy has dried, Dee holds 57577 at arm's length while she peels the duct tape off his bill and pulls the Qantas bag from his head. 57577 blinks in the sudden sun but is calm. Dee lets him go, and he gives her a quick, inscrutable look before shuffling back to his nest. Easing in front of his chicks, he examines this alien object stuck to his back. He nudges it with his bill and nibbles furiously at it, but the tag is glued fast.

We pack up the gear and start back to the house. "Don't forget to flag his nest," Dee says over her shoulder. Oh, yes. I tear off some green flagging and tie it next to the fading orange flag. 57577 stops nibbling to watch. Two months ago, he was just a penguin, one among hundreds of thousands, until we banded him for reasons I can't recall. He has since shown us some of his life, and now he will show us more. When he swims away he will sketch a line across the Atlantic while we follow from afar. A virtual creature, he will be the instrument of our revelation.

I heft the satellite backpack and trot after Dee, who is already far ahead of me. (It is gratifying to see that even her shadow has trouble keeping up with her.) Back at 223L, 57577 resumes his nibbling, then gives up and lowers himself to his belly. He settles in the dust and closes his eyes against the evening sun. He is still there the next day when I come to check on him, but the morning after that, he is gone.

9

Other People's Penguins

Back in September, spurred, I suspect, by my wrangle of 35472, Ginger made sure to teach us what she called "the tourist hold." Its purpose, she explained, was to obscure the fact that a viselike grip of the neck was necessary to handle a penguin. Tourists didn't like to see this. The apparent violence upset them. If we had to wrangle a penguin in front of people, then we were to grasp its neck with one hand while hugging its body with the other, as if we were cradling a large stuffed animal. Only then could we lug it off to a more private spot to do science on it. Ginger warned that the tourist hold wasn't as secure as a regular hold—various purple welts about my person would later attest to this—but it was important to keep up appearances. "You want the penguin to at least look comfortable," she said.

I am thinking of this now, of appearances and comfort and the prospect of another purple welt, as I haul a male penguin from his burrow. It is a hot afternoon in mid-January and I'm in the Doughnut area, near one of the elevated walkways on the tourist trail. We are usually discouraged from wrangling penguins so close to the trail, but I need to weigh this guy, and it's the first time we've seen him for several weeks. Who knows when I might see him again? So we tussle, he and I.

A gaggle of tourists has stopped on the walkway to watch. The penguin flops, snaps, struggles—a bit dramatically, I feel. Maybe he knows we have an audience. I finally get him in the tourist hold, and we are both panting when I hear, "Excuse me? Excuse me!" Standing over us is a dapper fellow in yellow shorts and a white polo shirt, with a pink sweater draped across his shoulders. "Are you hurting him?" he asks in English with a light French accent. "Excuse me, but it seems to me like you are hurting him."

"No, he's fine," I say, turning back to the penguin. It would look suspiciously furtive were I to spirit him off under my arm, so I decide to weigh him right here. More tourists have gathered. Someone has a camera out. The penguin kicks when I hoist him up, but the heat has sapped his fighting spirit. The scale settles—4.8 kilograms, or well over ten pounds—and I let him go. He scrambles back to his burrow and brays from its depths.

The show over, the crowd turns and scrunches away, but the fellow has remained, I see. He waves from the walkway as I collect my things, for he has had another thought. "Excuse me," he says again. "I am wondering, how do you justify all of this that you are doing?"

I eye him. He is irritatingly clean. His hair is nicely mussed, but also so fiercely pomaded that it doesn't budge in the wind. His cologne tastefully overpowers the prevailing scents of dust and guano. "Don't worry," I say as politely as I can. "Nothing we do can hurt these birds."

∾

Every year, tens of thousands of tourists flock to Punta Tombo. They come, they stand in line, they pay a fee, they pass through the gate. Once inside, their options are limited. The reserve may cover more than five hundred acres, and the colony is much larger still, but people aren't allowed to wander through it like we are. They must stay on the tourist trail, which is about half a mile long.

Along this short span they have their encounters. We've seen all manner of them: tourists using enormous telephoto lenses to take pictures of penguins less than three feet from them, tourists head-wagging at penguins, tourists flopping next to penguins on the ground while the penguins gaze at them quizzically. Some tourists hold out their shoes so the penguins can tug on their laces. A few hold out their fingers and need medical attention as a result. The mere prospect of seeing a penguin seems to elicit a jubilant insanity. I hear one story, which may or may not be true, of a pair of tourists who showed up in full marine survival suits.

A fairly typical summer afternoon at Punta Tombo. During the peak of the tourist season, five thousand people might visit the colony every day. The greatest crush comes when cruise ships dock at Puerto Madryn and send their passengers by tour bus to the colony. (Photo by Dee Boersma)

Penguins live only in the ice and snow, don't you know, and this couple wanted to be prepared.

Like the penguins, tourists came in a trickle at the start of the year, but now we are overrun. Unlike the penguins, the tourist population has been increasing steadily through time. In the 1960s, fewer than one thousand people made the long drive over rough roads from Trelew to the colony each season. Now that many of those roads are paved over with smooth and inviting blacktop, more than one hundred thousand people might come. In December and January, the months of peak visitation, the tourist trail can be thick with as many as five thousand people a day.

Speaking for myself, I tend to look askance at tourists. The arrangement is awkward for all of us. They expose my pretentions of working in an unpeopled wilderness, and I thwart their desire to visit one. El and I stick out in our shabby clothes with our mysterious tools, our pants smeared with guano and blood. We might as well be part of the exhibit, and for some tourists we are, our most mundane activities interesting by virtue of our presence. One day while El was washing her socks in a bucket next to the house, a tourist broke away from his guided group and came over to photograph her. She motioned for him to go away—the human equivalent of the head-wag—but he pointed his camera at her, so she very slowly and deliberately turned her back on him. She heard him storm off in a huff. How dare she not play along! "Now I know how the penguins feel," she said.

Even Turbo can be a bother when tourists are around. It isn't that he nuzzles up to strangers and ignores us so much as his ardor for us climbs to inconvenient peaks. He dashes out from his bush whenever he sees us, but we are to ignore him. The *guardafaunas* have decreed it. *Penguins are wild animals, not pets!* they have scolded us, and so I gently nudge Turbo aside when he tries to seduce my boots. "Not now!" I hiss as the tourists goggle, for they have been told in no uncertain terms that penguins are dangerous. Often, to avoid Turbo I have to run to the house. He chases me, clucking, but has yet to catch me. He is still a penguin, after all—slow on land.

꙳

Once all the tourists have left and the park is closed for the evening, I make a point of going back outside to visit Turbo. I feel for him. Where other penguins have progressed—finding mates, laying eggs, raising or losing chicks—he is still stuck in his own private limbo, struggling to satisfy impossible urges. I want him to know I haven't forgotten him, that I still care. He is quick to put me at ease, bless him, scooting out of his bush: *Helloooooo, friend!* I scratch his chin and rub his back, and we sit. Dusk falls, the moon rises, the penguins bray, Turbo and I are together, and some measure of rightness is restored, at least until tomorrow, when we will again have to pretend we don't know each other.

Tourists may be a nuisance for the two of us, but they seem to be less so for the penguins of Punta Tombo. Dee first studied the birds' reactions in 1989, with Pablo Yorio. Forty thousand people visited the colony that year, a number she and Pablo later wrote was "extraordinarily high." ("Little did we know," she says now.) To her surprise, and perhaps consternation, the penguins appeared to be fine with all the human traffic. After a brief period, they stopped bolting when people approached. Where reproductive success was concerned, their chicks were about the same size as those of penguins in the colony's more remote areas, and they fledged at similar rates.

"We suggest visitation may be compatible with penguin reproduction if visits are controlled," Pablo and Dee wrote in the subsequent paper. Their warning leavened reluctant acceptance; Dee didn't consider the issue settled at all. She knew there was more to a penguin's well-being than its outward appearance. Studies of other species have shown that even if a bird looks calm when a person walks close, its heart is racing as stress hormones course through its body.

She enlisted another graduate student, Brian Walker, to study the penguins' physiological response. Brian's first task was to translate a tourist's behavior into a protocol. With each penguin subject, he started by sitting quietly next to its nest. Then he walked around the nest and talked to himself. Then he sat next to the penguin and talked to himself. Then he walked around some more, still talking to himself. Then he crouched near the penguin and stared straight at it. After fifteen minutes of this, he wrangled the penguin and drew some blood from its foot to measure whether the concentration of corticosterone, a stress hormone, had increased at all during this performance.

Brian played the tourist for one group of penguins in a heavily visited area and another group more than a mile away. As bizarre as his behavior seems to me, the birds in the tourist area weren't terribly bothered by it; they reacted to him much more mildly than the penguins in the distant area. Also, the panicked response of those latter penguins dulled within about five days as they grew accustomed to Brian's antics. Brian cautioned that Magellanic penguins may experience long-term consequences from tourists that his short study couldn't capture, but they seem to know humans aren't out to get them.

This isn't to say nothing bad ever happens to penguins because of tourists, but incidents rarely rise above the level of ghastly anecdote. One that Dee tells from time to time is that, until a few years ago, the tourist trail was an actual road. The trail may be barely ten feet wide in places, but tourist buses used to lumber down it and park in a big graveled area. One year, the buses ran over a few penguins. Now, everyone has to walk.

"Maybe Punta Tombo is like Disney World," Dee mutters one afternoon while we wait for a stream of people to pass so we can cross the trail. Some penguins wait alongside us. Dee glances at them, all of us standing around, getting nothing done, wasting time. "Maybe we can keep packing people in and it won't matter," she says, sighing.

Regardless of the lack of strong, direct, negative effects, Dee has much to say about how the Argentine authorities handle tourists at Punta Tombo. She doesn't like how they have changed the nature of the colony from when she had run of the place in the early 1980s, when it was just her and her crew and hundreds of thousands of penguins. Tourists can be a force for good if well managed, but in her opinion the government lets in way too many. On top of that, they aren't charged enough. (The fees are tiered: Chubut residents can get in for three pesos, Argentine citizens for ten, and everyone else for thirty, or about ten US dollars.) The money collected from them doesn't stay with the penguins, necessarily, but can be shunted into the general coffers. "They're undervaluing the penguins," Dee grouses.

To encourage so many visitors initiates an all but irreversible cycle of development, for, as Dee says, "all these people need a place to eat and poop and pee." The province builds structures to accommodate those needs and consequently has to recoup the construction costs, which it does by letting in more people, who then need to eat and poop and pee, and off you go. So we have seen. Close to the field house, there is a small cafeteria and gift shop where, if you are so inclined, you can buy stuffed penguins, or postcards, or stickers, or mate gourds, or wooden salad forks carved to look like penguins. (Dee gave El and me a pair as a present.) Next to that is a large public restroom, newly expanded this year. A tanker truck rumbles down from town at least twice a week to deliver thousands of gallons of water so the toilets will flush. They might as well halve the trips: in the men's restroom more than half the toilets are backed up; El reports similar circumstances on the women's side.

In this built environment, we are beset with a host of nagging infrastructural insults. Stuff is always breaking down, leaking, shorting out. The fridge works or it doesn't, according to its humor. A construction crew dug a deep pit near our trailer for some reason, and a penguin fell in. (We rescued it unharmed.) The worst was last October, after a pipe burst next to the restrooms. Sewage bubbled up from the ground and ran past our trailer in a steady, stinking stream for days. The piss flooded a VIP penguin's nest, and the pair lost their eggs. Dee was duly disgusted on all our behalves. "I can't wait to show him this," she said.

"Him" was a bureaucrat from the Chubut Dirección de Turismo who was scheduled to visit the colony. I never learned his name, but he came the next day, poor fellow. On our way back from our morning's labors we saw him: a tall, erect man, broad-shouldered, with an elegant sweep of black hair. In his starched white shirt and crisp dark pants, he looked a little bit like a penguin. Dee marched right up to him. She was several inches shorter than he, but got right in his face, pointed this way and that. El and I stood off to the side, openly staring. Dee said "*no functionar*" several times. Her Spanish is largely self-taught, her syntax unorthodox. El asked Miguel, the head *guardafauna* who was watching with us, whether Dee could be hard to understand. Miguel smiled in a Dee-related way I was learning to recognize. "She makes herself clear," he said.

She certainly did that day. The government man recoiled in the face of her onslaught. When he was at last able to collect himself, he nodded administratively and scribbled in a handsome leather-bound notebook. After he left, I asked Dee how Argentines, and especially men in this famously macho culture, react when she dresses them down. She chuckled. "I'm a woman, so I don't think they know quite what to do with me," she said. "If I were a man, there's no way I could get away with this."

Dee is always jousting with the Argentine authorities, sometimes over their granting her access to the colony, sometimes over their plans to develop it further. As amusing as it can be when she throws her weight around, it is at times uncomfortable to watch. I know she has the penguins' best interests at heart and from her many years here no shortage of reasons to be skeptical of the government's intentions. (An aquarium? In the desert? When they can't even get the toilets to flush?) But during these moments, I feel most strongly that we are American scientists who have invited ourselves to Argentina, where we study an Argentine animal and lecture the Argentines on what they should do with it. No wonder they bristle when we tell them they are fools.

At heart the issue is one of stewardship, which is another way of saying ownership, I suppose. Whose penguins are these? The question forms the subtext for all the tiffs between Dee and the province. Trace the arrivals back, weigh the competing claims. Dee has studied these penguins since 1982. No one knows them better than she does. To me, to her colleagues all over the world, these are her penguins. But how did this come to be? That Dee is in Argentina at all is due to Bill Conway, another American, who in 1964 fell in love with Punta Tombo's rough wildness. It was Conway who helped persuade the Chubut government that the colony, which had been lightly protected since 1972, should be more rigorously studied (and, as a consequence, developed). From his efforts, the Punta Tombo Reserva Científica opened on December 4, 1979. He was there for the opening ceremony, a day full of pomp and circumstance. The provincial governor attended, along with a small military band. A priest walked among the penguins and sprinkled holy water on them. He intoned a prayer, asked God to watch over his flock. This is also a claim of a sort.

Also present was Luis La Regina, the owner of a large *estancia*, or sheep ranch, called La Perla. I imagine he felt ambivalent as he watched the proceedings: the penguin colony was—and still is—on what had been his land. A small souvenir photo book, the closest thing I can find to an official history, tells of the transfer in 1972. "As the amount of people coming to the estancia to see the penguins began to increase so much," reads the translated passage, "Luis Emiliano La Regina donated

210 hectares of his lands to the province of Chubut because he thought that the area should be protected properly." Dee tells me the account has been sanitized. "It was the 1970s, and there was a military dictatorship," she says. "The La Reginas didn't have a choice." A few months before, sixteen political prisoners had been executed in a Trelew prison, in what is known as the Trelew Massacre. Against a growing authoritarianism and its lethal appetites, who dared refuse a government request, even one about something so measly as a few hundred thousand penguins?

La Regina was forty-nine years old in 1979. He had been head of the *estancia* since 1963, when his own father, who was also named Luis, died at the age of eighty. Luis La Regina Sr. had immigrated to Argentina from Sicily in 1896, crossing the Atlantic alone when he was just thirteen years old. He found a job as a field hand near Mar del Plata in Buenos Aires Province, but after a few years moved south to Chubut. He worked at a couple of *estancias* until 1929, when he borrowed money to buy the land that he would call La Perla. He married and raised a family and eventually had his own flock of eight thousand Merino sheep.

It is largely because of sheep ranchers like Luis La Regina Sr. that penguins breed on the Argentine mainland in the first place. Before the

A portion of the La Regina's sheep flock on the berm that separates the colony from the beach. This was one of several reminders that the penguin reserve sits in the middle of a large, working sheep ranch.

ranchers came, there were too many large predators. *Colpeo* foxes and pumas ranged over the campo, and penguins, like other seabirds, kept to the safety of islands. To protect their flocks of sheep, *estancia* owners and their field crews shot or poisoned the foxes and pumas. In their absence, penguins began to move ashore. A blurred black-and-white photo from the early 1900s shows a few penguins standing on one of the nearby beaches. Ana La Regina, the elder La Regina's wife, recalled small groups of penguins nesting on the peninsula in the 1930s, before the colony grew and grew into, for the time being, the world's largest for this species.

Go back further, to this place's pre-penguined past, when history sheds its local details and gets swept up in larger geopolitical currents. Luis La Regina Sr. was among a wave of European immigrants whom the Argentine government encouraged to settle the Patagonian south in the late nineteenth century. The settlers were preceded by a military campaign called the *Conquista del Desierto*, or the Conquest of the Desert. Begun in the late 1870s, its purpose was to pacify the region's restive indigenous peoples and make clear that the land was Argentine above all. Soldiers killed more than one thousand indigenous people and drove fifteen thousand more from their homes. This ultimately led to the extinction of several of the smaller tribes.

The campaign lasted until 1884, when a final band of some three thousand rebels surrendered to government troops in what is now Chubut Province. Some of the rebels were from groups of nomadic hunter-gatherers known collectively as the Tehuelches. The Tehuelches had lived in southern Patagonia for more than ten thousand years before colonists arrived. The region's name is thought to come from them: *Patagon* means "big feet," and the Tehuelches were reported to be giants by the first Europeans to meet them.

Those first Europeans were Magellan, his crew, and his supernumerary scribe, Pigafetta. "But one day (without anyone expecting it) we saw a giant who was on the shore, quite naked, and who danced, leaped, and sang, and while he sang he threw sand and dust on his head," Pigafetta wrote three paragraphs after his description of the strange geese and goslings. "Our captain sent one of his men toward him, charging him to leap and sing like the other in order to reassure him and show him friendship. Which he did. . . . And he was so tall that the tallest of us only came up to his waist. . . . The captain named the people of this sort *Pathagoni*."

Magellan and his men spent several weeks in 1520 among the Tehuelches, trading and hunting with the giants, learning a little of their language, showing off guns and mirrors and other tools. Before the Captain-General left, he captured two young Tehuelche men and imprisoned them on his ship. He meant to take them back to the Spanish king as a prize, but they died soon after they were taken from their homeland.

∽

Intrusion and displacement, absence and erasure and atrocity. They layer this place. They define it. The *Tombo* of Punta Tombo is thought to mean "tomb." Whose tomb the name refers to is unknown, but on South Beach, near the start of the survey, there is a human skull. Concealed among the tufts of pampas grass, covered in shell and sand, it is worn, weathered, and toothless. Plants poke up through a hole in the parietal bone. Below the skull, someone has arranged a guanaco's leg bones into a roughly human shape. On one side, where the arm would be, is a large, heavy bone that can only have come from a whale. Rocks encircle the arrangement. The skull is small. It did not come from a giant. Perhaps a young person, or a child. Dee told me it was thousands of years old.

Is the land anything other than an ossuary? The dead are all around us, human and animal alike. Bodies everywhere we look, most recent, some not, but all of them lying where they fell, half-buried in the sand.

∽

Not all the dead are buried yet. It is a couple of weeks after the dapper fellow, and El and I are walking to the house at the end of a day. The reserve will close soon so only a few tourists remain in little, scattered bands. We are going through Sea-Tip when a man hails us from the tourist trail. He is older and heavyset, swarthy, friendly. He is here with his young sons, he tells El in Spanish, and they saw a penguin that doesn't look like it is doing so well. They want to know what is wrong with it. Would we mind if they showed it to us?

The man and his sons lead us up the trail and point to a spot out in the campo. A penguin is lying in a burrow, its head turned away from us. It isn't moving. El and I walk over. Closer, we can see he is a large male. Blood has pooled in his bill and is dribbling in the dirt. His eyes are open, but sunken and blank. Flies have found him. They cluster on his juicier parts.

El and I walk back to the man. "I'm sorry," El tells him. "The penguin is dead." "Oh," the man says, "that's too bad." He murmurs something to his sons and nudges them away. They gallop off toward the end of the trail. The man gives the dead penguin a final, regretful look, nods to us, and follows after them.

"We can't leave him here," El says when we are alone.

"We should probably necropsy him," I say, hefting the penguin by the base of a flipper. He thumps against my leg. He is in rigor mortis and so has been dead only a few hours. His body is stiff; his flippers stick out straight, as if he is trying to glide through the air. El and I scuttle to the house with the body, ducking behind bushes, keeping out of sight. If Dee doesn't want us to handle live penguins in plain view, then she most certainly doesn't want us to cut into a dead one in front of everyone.

We stash the penguin behind our trailer. While I put on gloves and get a scalpel ready, El readies the notebook devoted to necropsies, called, rather solemnly, The Book of the Dead. It is filled with measurements and notes from all the dead adult penguins we have found this year. (A dead adult is much more meaningful than a dead chick. Ecologically speaking, chicks are a dime a dozen, but adult survivorship for Magellanic penguins is close to 90 percent, so a dead one represents a significant loss of potential reproductive output.) Anything deceased goes in the book, from bodies ravaged by giant petrels to a single flipper we found on the berm with a band closed around it, meaning it must have come from a study bird. If we have time when we find a dead penguin and the cause of death isn't obvious, we are instructed to do a necropsy, but even though we have been here for four months, I have yet to do one. I never needed to, not when Ginger or Dee were here, and there were always people more eager than I to slice into a dead penguin.

El and I first measure and weigh the penguin, and then I place him on his back and kneel before him. How did Ginger and Dee do this? They started at the cloaca, I think, so I pinch a gobbet of loose flesh there, slide the scalpel in, and draw it through the feathers toward the sternum. The blade catches on feathers and gristle, and I have to tug it along. Blood starts to spill out. I cut the penguin's flesh all the way to his neck, and slice at the film that holds his dermis to his muscles and peel his skin away. It is like drawing back a curtain. His powerful breast muscles are the first thing I see. They run nearly the length of his torso and are almost an inch thick. Purple and dense, they remind me of tuna sashimi.

The penguin has no obvious wounds or injuries, but a lot of blood has bubbled on his breast plate and subdermally. We recall that he was lying outside his burrow with his flippers and feet splayed, which penguins do when they are trying to cool off. According to what Ginger and Dee have taught us, this means he died of heat stress. In essence, the day was so hot he cooked to death in his own skin. El writes "COD = heat stress" in The Book of the Dead.

We have established what we think killed the penguin, but I keep cutting, sawing through his breast bone and tugging it open. It yields— *crack!*—revealing the intricate gore of his internal machinery. Shades of light pink, red, blue, deep purple. Organs that until a few hours ago were tightly packed and pulsing have relaxed, relieved of duty. I don't know all their names, but I can tell the basic stuff: the shiny slab of liver, the deflated lungs. Ropes of his intestines slide through my fingers as I root around his body cavity. Blood sloshes on my gloves, onto his once white belly. His insides are so warm.

I find cutting into this penguin intoxicating. Everything connects to everything else. Pull on one thing, another dances—true in ecology, true in anatomy. Here is his heart, the same size and color of a ripe plum, and having the same firmness when I squeeze it. Here is his stomach, which I slice open with an easy *snick*. Out flows a gout of grayish slime, and I see

Cutting into a dead male penguin to try to determine how it died.

some portion of one of his final meals: one small clam shell and a squid beak. This gives me pause, thinking of the penguin gobbling up the empty clam shell as a calcium supplement, or snapping up that squid, his mind already on the next bite, on his maintenance and his nest and his future, unaware he will never eat again.

"That's probably enough for me," El says, standing up. She has made notes, but watched mostly in silence as I hacked away. She doesn't like the necropsies, not because she is squeamish, but because she thinks they are violent. A penguin is dead, and in our hunger for information we do violence to his body. She leaves to start dinner. "Come in when you're done," she says over her shoulder. "You can chop up some vegetables if you want."

Chastened by the rebuke, I lug the body off and dump it behind a big *duraznillo* bush, well away from where anyone might see it. The penguin's entrails stream out behind him in the dust. I do my best to tuck them back in, to give this bird back some semblance of order and dignity, but it is hopeless. He has been undone and there is no putting him together again.

Some kelp gulls have floated over to see what I'm up to. They circle above as I pile the penguin's guts next to him in what I know will be a very temporary shrine. When I walk away, the gulls land and start fighting over his bits and pieces. I leave him to them. He was his own for all his life, and then he was the project's for a few minutes, and now he will be theirs.

10
The Penguin Coast

Cool as she can sometimes be to our work, El has come to love its routine. She loves how we wake with the sun for the Cañada check, come back for breakfast, go out and work with the penguins all afternoon, check Cañada once more on the way home in the evening, make dinner, and, once we have said goodnight to Turbo, retire to our trailer as the sun goes down, usually to tuck into a dense Russian novel by head lamp. (El has lately been reading *Cancer Ward* by Solzhenitsyn; I, Dostoevsky's *The Brothers Karamazov.*) "Everything we do is right outside the door," she said one sun-drenched evening from our trailer's stoop. "It's really nice."

Our settling into this out-of-the-way life has been gradual but instinctive. Come to Punta Tombo for the penguins; stay for other reasons. I remember what a relief it used to be when we had an excuse to go to Trelew. Before we ransacked La Anonima, we might walk through the city plaza, or sit in a park, or get coffee, just to vanish in the civic anonymity only crowds of indifferent Argentines could provide. But Trelew has lost that allure, the city now a reminder of the urban hubbub we will return to in a few weeks. Anyway, gladiatorial shopping is a bore. These days, if we have to leave Punta Tombo at all, we like it when the protocol points us to the lonelier parts of Chubut.

We are pointed in such a way late one afternoon in January, to a stretch of coastline about a hundred miles south called Cabo Dos Bahias. We leave as soon as we finish weighing all the study chicks in Sea-Tip. (They are getting so big now!) Opus rumbles away from Tombo, and as I turn left on Ruta 1 instead of right, El's eyes start to droop. Her head lolls. Soon she is asleep. Moving vehicles are a powerful sedative for her.

I settle in for the three-hour drive. No other cars on the road. The sky is white, the sea a washed-out blue, the land evocatively rustic. I roll down the window and hold my hand out, pretend it's a wing (or a flipper). A telephone wire mirrors the movement, smoothly undulating along, up and down, up and down. I wonder where it is headed—there have been no structures for miles—when the poles suddenly stop and the wire's journey comes to an abrupt halt. It hangs from the last pole, flapping sullenly in the wind as it shrinks in the rearview mirror, and then Opus goes over a rise and I can't see it anymore.

⌁

We first learned of Cabo Dos Bahias when Ginger and El visited the colony back in October, a few days after Dee arrived. The penguins at Cabo Dos Bahias stand in counterpoint to those at Punta Tombo for Dee's purposes, and Ginger was off to deploy a few satellite tags. Dee told her to bring a field worker. "Whoever you want," she said. "I'll take El," Ginger said with an alacrity I didn't realize was insulting until later.

Ginger and El left the next morning. They were only gone for about thirty-six hours—they stayed overnight at a small resort that rents out boxcars refurbished with bunk beds—but it was the longest El and I had been apart since we had come to Argentina. When she and Ginger returned the following afternoon, I bounded over and pressed her for details.

"What was Cabo Dos Bahias like?"

"It was a penguin colony," El said, and shrugged. "There were a lot of penguins." She was tired after a couple of long days.

"Anything else of note?"

"I don't know. It was really windy."

"Oh. Okay."

"I'm going to take a nap."

We were neophytes then in this world of penguins. One penguin was another penguin was any other penguin. Our sensibilities have since become more refined. We are more attuned to nuances—the grays, if you will, in this world of black and white. When we pull into the entrance at the Cabo Dos Bahias Reserva Natural, we see a world subtly different from the one we know. The ground is rockier, hillier, the topography more diverse. Instead of long beaches open to the sea, there is a cove—a baylet between the *dos bahias*—and it is here the penguins gather to preen and bathe and size each other up. Most of all, the colony has only ten thousand pairs or so, and the penguins themselves don't dominate the tableau. I can hear a certain amount of braying, but the *hooAAAAAAAAAAAH*s are blended with larger fields of sound. I have a sense that more colonies in Argentina are like this one than Tombo: the penguins notable, but also somewhat modest.

The Magellanic penguin colony at Cabo Dos Bahias. This colony, with around ten thousand breeding pairs, is considerably smaller than the colony at Punta Tombo.

We check in with the sole *guardafauna*, a portly fellow who had forgotten we were coming but doesn't care either way. He waves us chummily through and returns to his lunch. We drive on to the parking lot. Opus is the only vehicle. We slip on our vests, grab our *ganchos*, and sling into the colony. The penguins we pass are clearly used to people—they don't flee at the mere sight of us—but they have kept their wilder edge. I feel them appraising me and am aware of being judged a large, blundering thing that is so far benign, but not to be trusted. This is probably a fair assessment.

Today, El and I have three things to do. We are to retrieve, if possible, some satellite tags Dee deployed here in December during Late Chick; we are to do a stake survey; and we are to weigh and measure thirty random chicks as part of the inter-colony random chick weigh, which sounds like it should have a monster truck rally at the end of it, or a pig roast.

We start with the stake survey. Back at Punta Tombo, Dee has spread short metal stakes along several transects across the colony. Some transects pass through bushy areas, others through more burrow-strewn sites. Earlier in October, and again this past week, El and I visited sixty-three of those stakes. Around each one, we tallied all the penguins, all the eggs (if it was October) or chicks (if it was January), and all the nests, be they active or inactive. By subsampling the colony's different habitats, the stake survey amounts to an annual census. Its principal finding—that the number of penguins has declined by about 40 percent since 1987—is the ballast for Dee's various arguments. Quibble with her if you must over this or that, but don't try to deny the steady, downward march of the line on her master graph.

The stake survey at Cabo Dos Bahias is the same as Punta Tombo's, but here there are only twenty-four stakes. (This seems atypically restrained for Dee.) We find the first in a patch of bare ground. I stand over the rebar and hold one end of a rope. El takes the other and walks out until it is taut. The rope is precisely 5.64 meters long, so she can pace a circle that covers precisely one hundred square meters. She starts a slow trudge, reading the landscape as she goes: "Active live bush three, one male . . . Burrow one, inactive . . . Live bush four, some use . . . Live bush three, active, two chicks . . ." (We distinguish between live and dead bushes. The numbers tell a nest's relative quality; one is best, five is worst.)

Prone to melancholy as I sometimes can be, the most poignant nests for me are the "inactive" and "some use" ones: those empty nests that have evidence of activity from this year and are now quiet records of spheniscid ambition and disappointment. A male may have claimed this bush and never gotten a mate, or a pair may have nested in that burrow but lost its eggs or chicks. Now they are all gone, leaving behind only a couple of sticks, a few feathers.

We finish the survey and turn to the random chicks, snatching up whoever happens to be near. The chicks wriggle and shriek, nip at us with their lengthening, strengthening bills. They stagger off when we let them go, stunned by our assault. They were not expecting this; they are not Tombo chicks.

Eyeballing their measurements, they are physically a little smaller than their northern kin, but to my grip they feel sturdier. Their bellies are rounder and tauter. The ground shows other differences. At Cabo Dos Bahias, the guano spatters have more of an orange hue, as opposed to the predominantly white spatters of Punta Tombo. In the semiotics of penguin excrement, this means the chicks' diet here is heavy with *Munida gregaria*, the gregarious squat lobster. (White means fish.) Squat lobsters are a small species of crustacean related to shrimp and crab. They can be an inch or two long, and they gather in dense shoals. The orange in the penguins' guano comes from the keratin in their shells. Judging by the thickness of the guano, the chicks must be dining very, very well on the squat lobster. The entrances to some nests are so heavily dolloped that, if I squint, I might be standing in a field of smashed kumquats.

⌐

Cabo Dos Bahias means "Cape Two Bays" in English. The two bays are Bahia Camarones to the north of the cape and the Golfo San Jorge to the south. The latter is one of Chubut's largest and more important bodies of water. About 50 percent of the province's population lives along the gulf's coast, which also has a substantial percentage of Argentina's oil and gas reserves. Bahia Camarones, while smaller, gets its name from one of its dominant fisheries (*camarones* is the Spanish word for "shrimp").

The cape and its two bays are significant in other ways. At just below 45° south latitude, they are the southernmost reach of the massive tidal front system that makes this part of the South American coast such a fruit-ful place for small fish. In this, they form a rough geographic boundary in the broader penguin diet. North of here, anchovy make up the penguins' preferred prey, especially around the Peninsula Valdés; to the south, Chil-ean sprat are more common. But here in this area of transition, conditions are more variable. Neither anchovy nor sprat occur at high densities, so penguins make due with a variety of prey: some anchovy, some sprat, but also silversides, squid, and the squat lobster.

While the squat lobsters the Cabo Dos Bahias birds seem to be feed-ing their chicks aren't as nutritious as fish, they are much closer to the colony. This might in the end matter more. Dee's satellite tags have shown that during the Late Chick period, parents from Cabo Dos Bahias swim a little less than forty miles per foraging trip. At Punta Tombo, penguins have to swim nearly seventy miles—the longest distance of any colony tracked in Argentina.

We speak often of a penguin's faithfulness to its home. We speak of how they come back to their natal colony as adults, how they use the same bur-row or bush year after year, how they might even nest within a few yards of where they hatched as a chick. It is a nice account, but incomplete. When they are young, penguins are quick to roam, to wander. They do so from the time they can first move on their own. Once their parents leave them alone in the nest, some chicks abandon it almost immediately in favor of another that feels safer. Several chicks might cram together in an especially good spot, ten or a dozen eyes or more shining back at us from the dark when we peek in and try to pick out the banded study chick from the rabble of peers.

"Chicks vote with their feet," Dee has said. Later, some will vote with their flippers. Over the years, Dee has found a few stray banded penguins nesting at Cabo Dos Bahias or other colonies north of Punta Tombo. Ear-lier this season, she and some of her Argentine students spent a couple of days at a colony near Estancia San Lorenzo on Peninsula Valdés. They did nothing but *click* penguins for ten hours a day, looking for banded birds. They *click*ed more than twenty thousand birds, and from all those found two that Dee had banded years ago at Punta Tombo. Dee was ecstatic. "That was time very well spent!" she said when she got back.

The system is one of great, slow movements. Most of what is known about the Magellanic penguin, and certainly everything I know, comes from the penguins of Punta Tombo, but Cabo Dos Bahias shows me the limits of my vision. While the number of penguins at Punta Tombo has declined, the colony at Cabo Dos Bahias seems stable. The penguins go up in places and down in others, they settle and abandon. What does this mean for the species in Argentina? The country has sixty-six known colonies. They range in size from new gatherings with fewer than one hundred breeding pairs to established colonies like Punta Tombo, with its roughly two hundred thousand breeding pairs. Of the colonies where such information is available, thirty-eight are growing and eight are shrinking. Some colonies, especially the smaller ones, might more than double in size from one year to the next, but these are not the only colonies to grow. The one at San Lorenzo, which is not even forty years old, already has nearly one hundred and fifty thousand breeding pairs and is still growing rapidly. If that trend holds, it will soon supplant Punta Tombo as the world's largest colony of Magellanic penguins.

These increases and declines don't seem to be explained by birth or death alone. It is not just that penguins at successful colonies raise more chicks than those at shrinking colonies, although this certainly has a lot to do with it. The data also suggest more than a few chicks must be leaving their natal colonies and settling somewhere else when they start trying to breed at four or five years of age. They may do some prospecting when they are juveniles or young adults, visiting different colonies, getting a sense of things, but their ultimate criteria remain a mystery. "We don't know how they make their decision," Dee has said. "Maybe they look around, they see a lot of fat chicks, they think, *This is the place for me*."

⌐⌐

El and I finish all our work by late afternoon and start back to Punta Tombo. Halfway home, we pass an abandoned settlement. A faded sign tells its faded name: Cabo Raso. I've read about it; it usually gets a few words in the tourist guides. It has been a ghost town since the late 1980s, when its last elderly resident died. We've driven through it before, but now we pull off to explore the ruins. The wind plays among the dilapidated houses. A sagging concrete bunker sits on the outskirts of town, covered with grass and graffiti. Near it is a small cemetery; some of the tomb lids are eerily ajar.

We make our way to an eroding bank, scramble over it to sit at a sheltered crescent of beach. Moments like this are rare: nothing to do, no place to be, at least for a few hours. The weather is sublime. "Let's just stay here," El says, closing her eyes and working her hands into the warm stones.

Farther down the beach, a single penguin is settled on its belly close to the water. It is small, a female. She starts to her feet when she sees us. We are surprised to see each other in a place all our respective relations have left behind. People were here and now they are not. Likewise the penguins move and settle as they will. The waves crash against the shore.

We shrink ourselves so as not to bother this penguin, but we have spoiled whatever solitude she was enjoying. She waddles down to the water, wades in, starts swimming away. When she is a short distance from the beach, she gives a drawn-out *waaaaah*. She repeats it, deep and solemn. *Waaaaah. Waaaaah.* She looks back at us. *Waaaah.* It is her contact call. She is hailing any other penguin within earshot to see if they might want to band together. She swims farther out into the surf, calling all the way to whomever might be near, until the waves drown her out. Maybe she was a failed breeder, or maybe she was young. Maybe she will find another penguin to join her, or maybe she won't.

I take off my boots and lean back on my elbows, bury my bare feet in the gravel. We spend almost all of our time obsessing over how penguins leave a colony, or how this leaving might result in colonies, and one colony in particular, withering away. I wonder what it might feel like to be the first penguin to decide to nest somewhere, to separate oneself from the flocks. To be a pioneer isn't easy in this world.

The newest colony of Magellanic penguins in Argentina that anyone knows of is called El Pedral. It is on Punta Ninfas, at the southern tip of Golfo Nuevo, near Peninsula Valdés. Penguins will show up almost anywhere on the coast, as we have seen, but a few years ago seven pairs of penguins decided to try to breed on that little bit of land. No one knows why. A few males came ashore early in September, found acceptable bushes and burrows, and then seven females came and laid eggs. They were all most likely young birds, and inexperienced. All the eggs were lost.

The next year, thirteen pairs came. Again, the males established nests, and again the females laid eggs. This time a few pairs managed to keep their eggs long enough for chicks to hatch. But, as before, they were inexperienced breeders, and none of the chicks survived more than a few days.

By this time, local fishermen had heard about the new colony. Although there were just thirteen pairs, the fishermen were less than thrilled at the idea of another penguin colony so near to their fishing grounds. They think penguins compete with them for anchovy. During the winter, they set fire to the bushes under which the penguins had nested, burning them to the ground. (So: penguins are not universally beloved.)

The next year, twenty-two pairs of penguins went to El Pedral. To protect them, Pablo Garcia Borboroglu, one of Dee's former students, put up signs saying the area was off-limits due to a research project he was about to start. Popi, as we all know him, is an inveterately cheerful man and a researcher with Argentina's national research council. He said he would model his methods at El Pedral after Dee's at Punta Tombo and come back year after year to watch the colony grow. Granted, the schedule would be a little less rigorous, at least in the early years, but he hoped to get busier and busier as time went on. The signs were good. That season, he watched as the penguins at this newest colony fledged their first chicks.

11
Passages

On a breezy dawn in the last days of January, I am sitting alone on the red rocks that overlook the beach at the end of the tourist trail. There is a nice perch here, a seat almost amidst the stones, and this shelters me from the crisp wind. Below, a giant petrel is scavenging—savaging, really—a dead penguin. Living penguins stand in a wide circle around it, so that the petrel seems to be giving a lecture of some sort, perhaps on the gross anatomy of the Spheniscidae. It grabs the dead penguin and shakes it violently. Whatever sinews held the carcass together snap, and pieces spray out. The petrel gobbles them up. The other penguins take a few steps back.

I search for something else to look at and decide to take a tally of the cuts and scabs my hands have accumulated. A few are from the bushes' ubiquitous thorns, and two or three are from the more memorable engagements with adult penguins, but most, and certainly the grisliest, are from the chicks. I underestimated them, I guess. They had seemed so small, and then they weren't. I remember the first one that bit me. I was holding it with one hand while reaching for my scale with the other, just as I had many times before, only this time the chick swiveled around and gouged a divot out of my

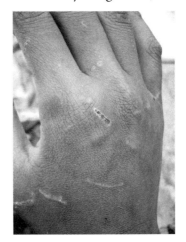

The scars from a season's worth of chick bites.

thumb. "Ow!" I yelled. "Dammit!" The chick was thereafter calm, having made its point.

That was several weeks ago. Scars now cover my arms, my hands. The most impressive are on the soft flesh between my forefingers and thumbs, two-inch-long seams, angry and pink. I'd go to El for sympathy, but her hands look worse. She gets bitten more than I do—she tends to hold the chicks closer and less tightly—and her skin is darker than mine, so her scars are more vivid.

I settle in my rocky seat, examine the penguins on the beach, check my watch (two minutes to go), reflect a little. Before El and I came to Punta Tombo, we were warned to take special care with the adults. Lose focus with them for a second, and they would flay our fingers. What I mean is we were expecting the penguins to hurt us one way, but they've left their marks in another. I have to hand it to them. Crafty beasts, those penguins.

⌒

I have tried, with mixed success, not to begrudge the chicks their defenses, even as I am often left bloody and grumbling. Other than that, watching the chicks develop has been a pleasure. After the initial blind helplessness, they went through what we called the awkward teenage years. All the indignities they endured! They stumbled around on feet as big as clown shoes. Their flippers hung at their sides like the sleeves of oversized dress shirts. Their down was often coated with guano, from the aimless asses of their parents, their siblings, the occasional passerby.

It took them time, but they grew into their bodies. Their movements became more coordinated. Floppy flippers stiffened with bone. Their bellies lightened, their faces whitened. Lately, their gray-brown down has started to slough off, replaced by the soft midnight blue of their juvenile plumage. Happily, they have also stopped dying; or, rather, dying as a matter of course. Predators might kill one every now and then, but that is a rare event we can ascribe to stochasticity, which is a scientist's way of saying bad luck. With their greater size has come bravery, or at least a newly defiant air. Some of the chicks, instead of running when I arrive to

This sequence shows chicks as they age, from one day after they hatch until just before they leave the nest. The entire chick-rearing phase, from hatching to fledging, lasts between 60 and 120 days.

measure them, plant themselves in front of me, daring me to take another step. Others seem to bray, although their bray is thin and reedy. One even leapt at my face, its bill clacking wildly, and bit me under the eye. We were both stunned by this.

Most of the chicks now weigh five or six pounds. Some look beefier than their parents. We started banding them once they weighed about four pounds. That is how Bartleby from nest F03B became 59785, Ripley from 100R became 59769, and Vaughn from V01 became 59543. It was bittersweet, helping them trade their informal identity for their official, scientific one. In so doing we bequeath them to Dee, to the Database—our small contribution to her project.

In these past weeks, we have started to sense the chicks' impatience with the land, their eagerness to leave it. They stand in front of their nests, flapping to build up their chest muscles. They start their calisthenics slowly, speeding up until their flippers are a blur. Almost inevitably there is a hitch of some sort—a pebble, an ill-timed gust of wind. They falter, totter, tumble over, carried away by their own momentum. After taking a moment to collect themselves, they catch their breath and begin again.

El and I walk through a campo alive with these windmilling chicks. Recently, we were talking about how it had been a couple of months since the last colony-wide phase shift. The earlier ones conceptually were matters of addition: from no penguins to male penguins, male penguins to female penguins, female penguins to eggs, eggs to chicks. Here we faced the prospect of a different phenomenon when the chicks depart: subtraction. Except El didn't like that word for obvious reasons, so we eventually settled on a better one.

⌒

The beach is thick with penguins at this early hour, thousands of them spread all along the coastline. Many more stream into or out of the colony proper. Dee calls this "the suits' morning commute," and they do seem to have a more corporate air about them, pottering along in dutiful lines as if they've just left a subway station. They waddle to the beach by way of

A view of North Beach during one of the early morning fledgling counts. Note the lines of penguins trickling down to the shore on the so-called penguin highways.

their preferred paths, known as penguin highways. Some head straight for the sea, but many stop to rest, or socialize with other penguins, or just rinse off in the surf and take a drink. (Like all seabirds, penguins can drink seawater, extracting the salt with a special gland just above their eyes.)

In each hand I hold a clicker. For the last fifteen minutes or so, I have been *click*ing penguins every five minutes, in two thirty-second samples each, as they pass through the nearest highway exit. Most are outbound: fourteen going out, one going back; eight out, two back. I will do this for an hour. It is relaxing work, one of the few truly restful tasks we have had all season. Every other day since the tenth of January, someone has come out here at sunrise to *click* adults for an hour. We do this so we can mark when the chicks start to fledge, but we have yet to see one. I check my watch: time for a count. Thirteen out, three back; eleven out, two back. The minutes pass. The wind is brisk against the sun's creeping warmth. My mind starts to wander, but I catch myself in time and start the count: ten out, two back; nine out, one ba—

Then, oh my heavens, I see a chick. It is so unexpected my mind doesn't register it at first, but there it is, a small thing, blue instead of black, passing over the threshold from the colony to the beach. It totters to the shore through packs of adults. Where they look somewhat jaded, the chick is wild-eyed, as if baffled by its own daring. Its presence seems to annoy the adults. They snap at it whenever it nears. The chick dodges out of the way until it at last reaches the ocean's edge. Here it stops. Waves beat on the shore, rushing up the beach before withdrawing with a caressing *hissss*.

The chick stands in place. It seems to have lost its nerve. It takes a tentative step, a wave hits, it scrambles back, the wave reaches for it, can't quite get it, retreats hissing. The chick looks relieved, but the next wave is bigger and travels farther up the beach. It upends the reluctant chick. Unable to resist now, the chick turns, drops its head, and rushes to meet the sea. There is a frothy jumble—roiling foam, a flipper flinging through the air—before the chick pops up like a cork in the calm water just beyond the surf. It looks both terrified and exhilarated as, for the first time, it immerses itself in the place where it will spend most of its life. It takes a few clumsy strokes with its flippers, reminding me of a tiny windup toy

A chick that has left its nest for good, about to pass under one of the tourist bridges on its way to the beach.

in a giant bathtub. But then it becomes aware of a natural and hitherto untested competence. It swims farther out, more sure of itself, past the scrum of bathing adults. Soon it is alone, a dot just visible in an endless plane of blue that sparkles with the fire of the spreading day. It ducks under the water, flicks its flippers, and disappears.

The word *fledge* comes from an old Dutch word, *vlug* (pronounced *vlooj*, I believe), which means quick or agile. In the mid-sixteenth century, the English took *vlug* and adapted it to its present definition, "ready to fly." So far as I know there is no equivalent term for a flightless bird that swims away from land for the first and most likely last time in its life. After watching that chick, I think there should be.

Fledglings show up more frequently on our counts after that. Sometimes there are so many we have trouble keeping track of them. One morning El counts thirty-eight, on another forty-two. Dee told us it would be like this. "You'll see one, and then it will just pick up and up and up," she said. "Like they can't get out of here fast enough."

El and I have been tossing theories around about how the chicks decide to leave their nest when they do, what they feel the moment they walk away. Their parents might stop feeding them, and hunger drives them off, or it might be the other way: they have one last, tremendous bellyful of food and decide to take their chances in the hope that fish might still be closer to the colony. Or perhaps they sense the coming change in the seasons through the shorter days. Really, we don't know how the sea calls to them, or why they choose to heed its call.

I don't know if it is this inexplicability, the awe it evinces, or something else, but my elation on seeing a chick fledge never lessens. When I walk back to the house after a count, my eyes are often wet with tears.

"Have you been crying?" El asked me once.

"It was the wind," I said.

"Sure it was," she said and nudged me in the ribs.

⌁

As we have watched chicks leave from the colony on our mornings, we also note the increasing absence of the chicks from the study nests, although we have yet to catch any of them in the act of fledging. It is probably better that we don't; if we did, we might burst with happiness. What typically happens is we get to a nest and find it empty, or, if not empty, then with both parents resting and looking especially contented. This is its own consolation. It starts to give the season a sense of an ending.

Sometimes I take a few seconds to look down the nest page at all the chicks' measurements, neatly boxed, from all our visits. Ecology, as I've learned all too well, can be deeply unforgiving, and those chicks that survive to fledge tend to follow a set trajectory: they hatch and start gaining weight until they reach a certain age and size, and then they leave. Nice and neat, start to finish, at least on paper.

Not all chicks followed that trajectory. Some had a sibling that died; those chicks always seemed to fare especially well. A not insignificant minority lost weight between our visits, only to regain it later. Even more rarely, we have seen second chicks that spent weeks substantially smaller than their older siblings. Most of them are dead now, but a few managed to hang on to life, their parents feeding them just enough so they didn't starve while they blithely stuffed the other chick full of food. Once the bigger sibling fledged, the parents turned their provisioning affections to their waifish second chick. It gained weight rapidly and fledged a week or two later.

So there are discrepancies, deviations. Ecology might not make much of these chicks' chances, but they are the ones I have been quietly rooting for. I suppose I'll always have a soft spot for the underdogs and weirdos of Punta Tombo.

⌁

One nest in particular comes to mind, in the Bungalows area: 404C. If the stories of most fledged chicks were straightforward, then those of the vast majority of nests certainly were. A male arrived, a female arrived, she laid eggs, etc. But some nests have stories that were trickier to unravel, and 404C is perhaps the most confusing of all.

404C began the year as what we call a parenthesis to another nest, 404B. When we were first checking nests in September, we also visited any nearby burrow or bush where a banded bird had been seen the year before. These sightings were listed in parentheses on the nest page. They weren't official nests, but that a banded penguin was there a few times was sufficient to merit a quick jaunt out of our way. Not that the banded penguin necessarily ended up nesting in the parenthetical burrow or bush. Often it didn't, and we stopped checking the parenthesis after a few weeks, but once in a while the penguin would get a mate, the pair would have eggs, and the parenthesis would become an official nest.

Such was the case with 404C, a forgettable bowl at the edge of a large *jume* bush thirty-three meters east of 404B. A known-age female, band number 45725, was seen there by herself at least once last season. When we first checked the parenthesis on September 17th, an unbanded male was there. He was alone for a month, and then one day 45725 joined him. She laid two eggs over the next four days, so we measured and marked up the clutch and toe-tagged her mate. He was now t2459. Lastly, we turned the parenthesis into a nest, 404C, just as we had many other times in many other places.

Here, events took their first odd turn. When El and Emily went back two days later, 45725 was gone. The toe-tagged male was incubating their clutch in her place. Her absence was regrettable, but also understandable. She had been light when she laid her eggs—less than eight pounds—and so was close to needing food. Her mate gamely incubated the clutch for nearly three weeks until November 10th, when 45725 came back. She had done well for herself, weighing more than ten pounds. Alas, an armadillo had eaten one of the eggs just before she returned, but the second egg remained intact, so we kept checking the nest.

At this point, 404C moved from odd to absurd. On November 16th, an unbanded female was incubating the egg. An unbanded, un-toe-tagged male was standing in front of the nest. A few days later, when we started visiting the nest every day so we could measure the chick when it hatched, the un-toe-tagged male was incubating the egg. He and the unbanded female shared incubating duties for the next few days, but they didn't seem to care about the egg that much. Why should they? It wasn't theirs. Sometimes, they left it unattended while they basked in the sun. The poor egg was little more than kelp gull bait.

On November 30th, t2459 was at 404C. We stood before him, confused. Had he come to reclaim his nest and egg? Why wasn't 45725 with him? Where had the unbanded pair gone? They were back the next day, it transpired, both of them bloodied from fighting. Had they battled with 45725 and t2459, who were nowhere to be found, and reclaimed 404C? In any event, they resumed incubating the egg, but by then the egg was at least forty-six days old, so we stopped coming to the nest every day. Since the chick hadn't hatched, we assumed the egg had been addled during the pairs' fight.

Nine days later, on December 12th, El and I visited 404C as part of a regular check of the Bungalows area. I don't know what we were expecting, but not what we saw: a new chick lying on its back in the rear of the nest. It peeped hoarsely and kicked its feet in the air. "Oh, no," I said.

An un-toe-tagged male stood in front of the nest. When El kneeled to retrieve the chick, the male scampered away and didn't look back. El put down her egg cup and reached into the nest and picked the chick up with her hand. "It's so light," she said. "I can feel its skeleton." She cradled the chick in her palms. "Oh, you poor, poor thing," she said, tears welling in her eyes.

We weighed and measured the chick and returned it. It was so weak that it couldn't prop itself on its belly, but tumbled onto its back and started peeping again.

Whenever we came upon a chick that was almost assuredly doomed, we would name it after a constellation or a star, like Orion or Vega or something. This chick being one of the last to hatch, we were running low on heavenly bodies—my recall of celestial nomenclature is limited—so we decided to name it Ceres, after the dwarf planet in the asteroid belt.

I'm not saying we didn't know the practice was silly, but it helped us feel better during those awful days, when so many chicks were dying and we could do nothing to help them.

By then it was the thick of December. We were in the busiest part of the season, measuring scads of chicks, alive and dead alike. There was tragedy to spare, so Ceres and 404C were quickly forgotten.

On the 22nd of December, El went to check the Bungalows area alone. 404C was one of the last nests on the circuit, and as El walked up to it, she steeled herself to find Ceres dead, or maybe not find him at all, a kelp gull having made off with his body. One more nest for the Ziplocs of Failure.

The unbanded female was brooding the chick.

El was so surprised she forgot to measure Ceres. We had to go back to 404C the next day on our way home for lunch. This time, the unbanded, un-toe-tagged male was there, having traded places with his partner. Ceres weighed more than half a pound and had been recently fed. He was, as Dee would have said, tight as a tick.

"He's going to live! He's going to live!" El cheered while I wrote Ceres's measurements on the nest page.

"He *might* live," I said. I didn't want to get our hopes up. "His outlook has certainly improved somewhat."

El turned to me and said very seriously, "We can't name him Ceres anymore."

"What should we do?" I asked. We might have been somewhat irreverent in our naming, but it was still a sacred act. We had never, ever changed a chick's name.

"We can add a name to it," El said. She thought for a second. "Let's call him Ceres Bucket." And so Ceres Bucket he became.

"What do you think is going on?" El asked when we were back at the house making lunch. The nest page was on the table between us. We were puzzling over it, trying to piece events together. The egg and therefore the chick had to be 45725's, of that we were sure. After that, nothing made sense.

Dee was eating a banana and looking over our shoulder. "Who knows?" she said with a shrug. "Just keep following the nest and see what happens."

So we did. Nine days later, on January 1st, Ceres Bucket weighed a pound and a half. He could stand on his own. When we came back ten days later, on the 11th, he weighed more than four pounds, and we gave him a toe tag. Ten days after that, he was big enough for the unbanded pair to start leaving him alone at the nest. Ten days after that, on January 31st, we started to see small blue feathers sprouting out from under his down. He weighed seven and a half pounds that day, so we banded him. We tried to choose a good band from the few we had left, and settled on 60123. We hoped these digits would bring him good luck.

⌐⌐

Through all of this, I sometimes thought back to Ceres Bucket's foster parents. A conventional ecological wisdom augured for their chick's death at so many forks: when 45725 left after she laid her clutch; when the first egg was lost; when the two pairs of penguins fought over the nest; when the unbanded pair claimed 404C, but only halfheartedly incubated the egg; when the male couldn't even be bothered to brood the chick after it hatched; and there were probably other circumstances we weren't there to see. Yet the unbanded pair had ultimately raised Ceres Bucket, and done so well.

The penguin literature didn't have much to offer. Adoption isn't a widespread phenomenon in seabirds, so only a few papers have been written on it. I read that little penguins in Australia will adopt a chick 1 or 2 percent of the time. In another study of emperor penguins, 10 percent of chicks were adopted (or abducted) by adults, but these adoptions usually lasted less than two days; longer-term adoptions were rare. One study of king penguins found that up to 25 percent of adults will feed an unrelated chick at some point during the breeding season, but that is hardly the same as adopting it. A king penguin even brooded a skua chick for an hour or so, before researchers intervened and gave the chick back to its frantic parents.

None of this seemed to apply to Ceres Bucket. With little penguin adoptions, the chick was always old and mobile enough to wander into the foster nest on its own. Emperor and king penguins that adopted chicks tended to be either failed breeders or unmated birds. Scientists think their behavior is an aftershock of frustrated reproductive hormones. Furthermore, unlike Magellanic penguins, both the emperor and king penguins stand together in huge scrums when they breed, as opposed to having discrete nests in bushes or burrows. It is easier for their chicks to move around and get mixed up.

Dee said adoption is almost unheard of with Magellanic penguins. She knew of fewer than ten instances in all her years at Punta Tombo. Where did that leave us? We didn't know what to do with good fortune, or dumb luck, or whatever it was that kept Ceres Bucket alive. We had no experience with it. We definitely weren't used to acts of what looked like selflessness or altruism among penguins, regardless of whether their motives were endocrinological or caused by a blown fuse somewhere in their peanut-sized brains. All I could do was add the unbanded pair of 404C to my roster of batty penguins.

We never saw them again after January 11th. Or, I should say, not that we know of. But every so often at the beach, when I watched some wet, shining adult dutifully waddle up from the ocean to the colony (sixteen out, one back), heavy with food for a chick, I would wonder if that bird was on its way to Ceres Bucket.

⌣

El and I followed Ceres Bucket for another couple of ten-day cycles. On one of them, we happened across 45725 and t2459 in a bush six meters west-southwest of 404C. We made a note to put the bush as a parenthesis on the 404C nest page when we prepared one for the upcoming season. Maybe the pair will nest there, maybe not. Let next year's crew sort them out.

The last time we saw Ceres Bucket, he was enormous. When we walked up on him, he was flapping in front of his nest. He weighed over eight pounds. His belly was so full he looked like he had eaten a bowling ball. He was blue and beautiful and strong. El was overcome. She picked him up and held him under his flippers as if he were a child. Ceres Bucket hung coolly in the air while she nuzzled his back, burying her face in his new feathers. "Look at him!" she said. "Look at how big he is!" She put Ceres Bucket down and went to pat him on the head. He nipped at her lightly, and she laughed. She was the happiest I'd seen her in a long time.

Some days later, we took a little detour on our way back to the house to check on 404C. The nest was empty. Ceres Bucket had left.

"Looks like he fledged," I said. "Whodathunkit?"

That was the last time El cried.

El holding Ceres Bucket during one of our last visits.

12

The Giant Petrel

It is February now, and the days are shortening, the high heat of summer laced with the coming autumnal cool. Our time at the colony likewise rushes to its close, and with that has come a series of summary assignments. One is the third and last oiled bird beach survey. Since we are just two now, El and I decide to spread it out over a few days. We do North Beach together, but when it is time to do South Beach the next day, El says she would rather stay at the house to do some cleaning, and so it is that late one balmy afternoon, I strike off to *click* all the penguins alone.

South Beach is one of my favorite areas in the colony. It is tucked behind some hills, so I can't see or hear the main road. The buses and cars and tourists and restrooms and the hamburger stand and the house and the *cueva* and the trailers and everything else—I can pretend none of it is there. The penguins of South Beach are also less habituated to people. When I approach, they stand up tall, curious but wary. I get closer and they hurtle away, tumbling over one another. They spread out in panicked waves as I walk across the campo, until I take pity and pick my route with greater care.

Up ahead, at the crest of the berm that leads down to the beach, the fleeing penguins have come to an abrupt stop. Peering over the berm and then back at me, they seem to be weighing one fate against another. It is not until I join them that I see why: eight southern giant petrels are resting on the sand below, arrayed, I note with some dismay, over the very spot where the survey is to start.

The penguins and I stand together on the berm, fidgeting in a discomfited little group. Given the choice between the giant petrels and me, the penguins will take their chances with me. I can't say I blame them.

Penguins on South Beach.

The first time I saw a southern giant petrel was in September. I was out on some undignified toil or another in the Factura area, at a nest near the berm. Lying on my belly, my face jammed in a burrow, I was trying to persuade a male penguin to shift so I could read his band number. He was of course having none of it. As the penguin and I huffed oaths at each other, I saw at the edge of vision a shadow racing across the ground. The shadow was so big that I thought for a moment, absurdly, it must be a plane. Then I made the shadow's source right before a giant petrel swept over my head, low and fast. The bird could not have been more than a few feet above me. I could hear the wind tear at its feathers and rush over its body. Already deep in its burrow, the penguin pushed back against the rear wall. Just as instinctively, I ducked under an adjacent bush. The giant petrel soared on, rose some thirty feet in the air, and then dropped down again almost to the ground, rose, dropped once more. It was enormous, its wings so long and thin they seemed somehow distorted by the heat. It must have reached the spot where I was at the nadir of one of its swift descents. I'm not sure it even saw me.

After that, I saw giant petrels almost every day, either as they skimmed over the campo or floated in the waves just off the beach. A few wore the lighter plumage of adults, but most were dark brown juveniles. These were like flying shadows. The shape of their wings brought rather oddly to mind the curved steel blade of a big paper cutter in my hometown's little public library. The blade had always scared me, a monster from the synthetic wilderness of childhood. Once, catching me frittering with it, the librarian made me watch as she dispatched a sheaf of papers with a single chop to show off the blade's sharpness. Be careful, she had warned, or this thing will take your fingers right off.

⌒

Giant petrels belong to the order of seabirds called the Procellariiformes. The root of the name—*procella*—means "storm" or "tempest." Members of the order, which includes albatrosses, shearwaters, gadfly petrels, and the like, are famous for their elegant flight. The largest of them, the wandering and royal albatrosses, have, at more than twelve feet, the longest wingspan of any bird. On stiff wings locked in place by tendons in their shoulders, they can soar for hours, dipping and climbing with and against the winds, flapping rarely, if ever, sometimes for hundreds of miles.

As literary figures, too, albatrosses enjoy a certain prestige, but giant petrels are considerably less loved. The two species, the northern and the southern, belong to a separate family, the Procellariidae. In this they are like taxonomic exiles: the brutes, the fell albatrosses. They are smaller than albatrosses, having a wingspan of just seven feet, but they are stockier, with thicker necks and bulkier bills. Using those bills, they are enthusiastic scavengers, gathering by the dozens around large, rotting whale or seal carcasses, from which they are frequently seen emerging, their heads clotted with gore. The savage effect is heightened by the iris of a giant petrel's eye, which can be white or light blue. Against a face speckled brown and white, as if spattered with dried blood, this can give the bird a singularly blank and pitiless gaze.

Overland, giant petrels trade more in menace than in grace. It was clear the penguins were uniquely terrified of them. Penguins would chase away kelp gulls and lunge at armadillos, but they ran for their lives from the giant petrels, which cruised back and forth over the colony like great airborne ids, making no effort to conceal their lethal agendas. We would see them parading around on the beach, their wings held out wide to herd some unfortunate penguin they had isolated from the flock. When we sometimes found a penguin torn limb from limb, its pelt flipped inside out to expose the bloodstained circuitry of bone and sinew, we knew who was responsible. Even other predators were intimidated. Once, after necropsying an adult penguin, we left its body on the beach. First, the kelp gulls came to investigate. As they were settling in, a pair of skuas swooped down and shooed the gulls away. The skuas had the penguin to themselves for less than a minute before a giant petrel crash-landed between them, sending them sprawling into the air. It commenced dining, untroubled. Thereafter, the competitive hierarchy was clear.

In spite of such an obvious if macabre charisma, few scientists have been eager to sink themselves into the life of the giant petrel. What few studies there are come mostly from Antarctica. In Argentina, where the southern giant petrel is a threatened species, it is comparatively

Two southern giant petrels feast on a dead penguin, as the other penguins watch, giving them a wide berth.

unexamined, save for the occasional census of the country's few colonies. But over the past decade or so, biologists have started to appreciate more fully the giant petrel's role as a predator in the Patagonian Shelf's food web. Recent counts have shown that close to twenty-five hundred pairs nest in Argentine Patagonia alone. More birds forage in the surrounding seas, some coming from as far away as the sub-Antarctic islands.

The colonies closest to Punta Tombo are sixty or seventy miles to the south, in the Golfo San Jorge. One, on Isla Arce just off Cabo Dos Bahias, hosts a few hundred birds. The colonies are studies in the balance of communitarian spirit with innate surliness. Pairs of petrels are arrayed evenly about six feet apart, just out of range of those massive bills. Their nests are short mounds of mud, or scrapes lined with pebbles and grass. As with the penguin colony, the air is redolent with an excremental perfume. Unlike the penguins, a giant petrel colony is fairly quiet. Perhaps this is due to the space the birds know to give one another.

Giant petrels lay a single egg shortly after the pairs arrive, usually in September or October. The chick hatches in November, and both parents bring it food for the next few months, ranging hundreds or even thousands of miles on foraging trips. Males and females tend to hunt in different areas. Males keep to the coasts, where they maraud among the penguins, which, whether through predation or scavenging, make up close to 90 percent of their diet. Females, while also feeding heavily on penguin and seal, fly farther out from land, where they eat the squid and crustaceans near the water's surface.

Subsisting on penguin and other unfortunate creatures, a giant petrel chick grows slowly. It molts into its dark brown juvenile plumage in February. The parents continue to feed it until April, by which point they apparently tire of supporting a chick almost as large as they are. They then leave the colony for the winter, but the chicks don't immediately depart. Instead, they stay at the colony alone, sometimes for up to two months, until, obliged by a keen hunger, they stretch their wings in the wind and lift into the air for the first time, flying off to search for something to eat.

⌐

Sometimes when I came across a giant petrel's gruesome leavings, I would wonder if it is possible to love a giant petrel in the same way we love the penguins, complicated though that love may be. Or, if that is asking too much, whether years of close study might at least evince something closer to affection. People who work with giant petrels will speak to their virtues; they are apparently much easier to handle than penguins, for instance, and don't bite. But history has not been as kind, judging by the nicknames with which the giant petrel has been saddled: glutton, stinker, stinkpot. The latter two pay spiteful homage to the giant petrel's unique odor—a petrel essence, if you will. It is an unctuous smell, so sharp as to be an almost physical sensation. It persists long after a petrel is dead. Outside the *cueva*, someone many years ago placed a giant petrel skull they found on the beach. The feathers have long since rotted off and blown away, but the smell remains. To turn the skull over in my hands is to coat my fingers with its residues.

"Its appearance and habits are alike unprepossessing," Robert Cushman Murphy wrote of the giant petrel. He had seen them frequently during his time on the *Daisy* and referred to them as giant fulmars (*fulmar* meaning *foul gull*). "A bird of prey," he went on, "it is, nevertheless, ungainly and uncouth, lacking the beauty and dash which win admiration for even the most bloodthirsty of falcons and eagles."

Murphy would compare the giant petrel to a vulture, to ogres. He likened a flock of petrels running from him in fear to the "swine of the Gadarenes." He mocked their "supposed courtship," when males "hideously smeared with blood and grease" danced like stupid peacocks before bored females. Most galling for a Procellariiform, he questioned the giant petrel's flying prowess. "At sea," he wrote, "the Giant Fulmar is a 'stiff flyer,' showing to best advantage only in high winds. It is a far less agile and graceful bird than mollymauks of the same size, and it assumes particularly queer and awkward attitudes when descending to the ocean under the handicap of a light breeze." He was willing to concede that giant petrels could at least walk and run better than albatrosses, but what chance did they have in the wider human imagination if seabirds' greatest champion could find almost no redeeming qualities?

Murphy's scorn was strong, but it may have concealed a darker fear of the storm birds. Giant petrels were the scourge of sailors in the southern oceans, where they attacked anyone who fell overboard. There were accounts of sailors having their arms sliced to ribbons as they tried to defend themselves, and giant petrels were known to target the face and eyes with that glorious hook at the end of their "fossil carved ivory" bills. Murphy himself included a tale I would read more than once, of a boatswain who fell off the *HMS Erebus* during a survey of Antarctica in 1840. Giant petrels immediately set upon the poor man, "whirling over his head," slashing at him with their bills. So frenzied was their assault that the man drowned and sank before the ship could come about and save him.

For Murphy, the giant petrel was "scarcely more popular as a bird than a shark is as a fish." This in the end should give some idea of the class of animals with which the giant petrel was lumped, hatred and calumny being the price any creature has to pay for reminding us that, just like everything else, we, too, can be prey.

Slowly, I make my way down to the beach. The giant petrels are unperturbed at first, but they rise, unhurried, when I get closer. In the strong gusts, all they must do is unfurl their long wings and hold them out so the winds will carry them away. They lift off and glide just past the breakers before settling on the swell.

I start the survey as I have so many others, counting off paces, *click*ing the penguins as they cautiously retake the vacated beach, scribbling notes as circumstances require. The first couple of miles are slow and uneventful. I find a dead cormorant in the wrack at the tideline. It has an orange aluminum band on its stiff, dried leg, and I make a note of the band number so I can send it to the proper authorities before prying the band off and putting it in my pocket. I will give it to El to wear as a ring. I know she will love it.

The oiled bird survey being longer than a regular survey, this South Beach count takes me much farther down the beach than I usually go. I walk past a spot we call Whale Point, where several skulls from pilot whales lie scattered about, then continue south well beyond the Grand Cañada, close to the southern boundary of the La Regina's *estancia*. The penguins are not stacked as thickly as they were on counts during the blistering days at height of the season; dozens spread across the shore rather than hundreds. Before long these peter from a steady layer to staggered groups of only a few birds. After a while, all I see are those mysterious penguins who like to be alone, sunning themselves.

When there is little to count, a count is no longer a count, but a pleasant walk in the sunshine. As late afternoon turns to early evening, the light grows richer, warmer, with no sound but the *whish* of the waves and the pampas grass. I pace along the beach and am losing myself in the sleepy warmth when I hear over the gentle surf a strange, strangled sound. I think nothing of it—it is a trick of the wind, or the wheezing of my nose—until I hear the sound again. Out past the waves, twenty-five yards or so from the beach, a giant petrel is floating alone on the water. Strange. Giant petrels hiss and grunt and croak; they don't moan and wail. Then the water in front of the giant petrel ruffles and a penguin's head emerges, and I know the source of the awful call.

The giant petrel reaches down with its bill and grasps the penguin by the neck and forces its head back underwater. Satisfied that it has subdued its prey, the giant petrel turns to nibble on something near the penguin's tail. Probably it is trying to tear open that soft flesh near the cloaca, so it can get to the savory guts. At this, the penguin rears up, gasping and flailing. The giant petrel pushes it underwater again, adjusting its own posture so its full weight is settled on the penguin's back. It relaxes, rocks with the waves. It has a predator's terrible patience.

It occurs to me that I am seeing what one philosopher has called the universal metabolism. I think of it as the moment a being transforms into a thing. Such a change, whether it happens in seconds or is spread out over days or weeks, can be intensely upsetting to watch. There is no romance when one animal eats another, none of that peculiar air of sport that is the hallmark of most popular depictions of predation—the song-bird flees from the hawk, or the gazelle from the cheetah, the race won by whichever animal is the better athlete, whichever got the better start. Not here. Here there is only hunger and pain.

"Nothing exists of or for itself, but only in relation to other organisms," Murphy has said. With this in mind, I feel I owe something to the penguin, unbanded, otherwise unremarked upon, seen far down the beach, well outside the colony. I sit down to observe its last moments. It is the least I can do. It is also the most I can do.

⤳

The literary critic Frank Kermode once wrote that no human need is "more profound than to humanize the common death." Punta Tombo has definitely not lacked for common deaths. El and I spent the months reconciling ourselves to them as best we could, if not exactly humanizing them: the deaths of the adults we could see, whether baked in the heat or killed by the foxes; the deaths of the adults we could not see, entangled in fishing gear, or starved at sea; and, of course, the deaths of all the chicks, seen and unseen alike.

We have also been taught, or told ourselves, that the penguins' common deaths are somehow different. Death may be intrinsic to this system, and most chicks will die even in a good year, but our unspoken aim is that we shall have no more bad deaths. We hold to this because many times we can't help but feel the deaths could have been prevented. In this dream world, a lot of the dead chicks would have been fine were it not for the storms climate change sent down on them, or the fishermen who hauled up too many tons of fish and squid and deprived them of food while subsidizing the kelp gulls that ate more of them. The adults the *colpeo* fox killed would still be alive today if the La Reginas had not allowed the foxes to roam across the *estancia* in the first place. Those bad deaths give energy to our cause. Death is something we can fix.

This giant petrel and this penguin, on the other hand, exemplify the simple act of one animal killing another for food. Although it horrifies me to watch—and thank heavens El isn't here to see the penguin's protracted suffering, she would break into pieces—I know there is nothing wrong with this penguin's death. In an ecological sense, this is a good death. Good, because it is a sign the local giant petrel population is growing. Good, because this giant petrel is eating a natural food, and not some shiny plastic toggle it saw floating in the middle of the ocean. (One study found that, after penguins, plastics are what they eat the most in Argentine waters; due to an unfortunate interplay between chemistry and physiology, plastic smells like food to petrels.)

When Kermode wrote of common death, he was developing a theory for the way literature addresses the desire to find meaning in the brevity and broad meaninglessness of life. Such meaning, he thought, can often be found in the embrace of apocalyptic visions and imminent crises, such as the end times that always seem to herald the collapse of life as we know it. Since Kermode was writing in 1967, those crises tended to center around war and nuclear apocalypse. Now, as we start to reckon with the environmental consequences of this human-driven epoch we call the Anthropocene, his arguments have more explicit ecological overtones.

They also point to another, larger tension inherent to ecology. If we accept Kermode's argument that (ecological) crises can be used to give greater spiritual heft to our own individual lives, then we also have to face that ecology does not really have a great deal of interest in an individual creature, to say nothing of its suffering, existential or otherwise. As I walk around Punta Tombo and cleave to the richness of its many lives in all their bright details, as I spend months measuring those details down to the millimeter, I know that I am here to aid in their ultimate abstraction, as those details are added to the existing millions of data points.

This is the crux that comes when nothing is of itself valuable save for what it adds to larger epistemological structures and theses: eggs into clutch size into yearly comparisons of preseason foraging success; chicks into growth tables into selection pressure and food availability and climate models. In the "ecological vision," as writer J. M. Coetzee calls it, every being has a role within a vast, complex, and interwoven system, and it is the role, rather than the being playing the role, that matters. So long as a steady supply of beings are available to play the role—beings that might be found in abundance, for example, at the world's largest colony of Magellanic penguins—the individual that happens to be playing the role at any particular time is incidental, if not irrelevant. A concern for the fate of the species supersedes any closer attention to an individual's plight.

If I wanted to, accordingly, I could with a clear ecological vision watch a Southern Giant Petrel (which could be any southern giant petrel) drown a Magellanic Penguin (which could be any Magellanic penguin), and add them to the annual crop of observations that will help us sketch the ecology of the Patagonian Shelf. Such a vision has troublesome implications. "An ecological philosophy that tells us to live side by side with other creatures justifies itself by appealing to an idea, an idea of a higher order than any living creature," Coetzee writes. "An idea, finally—and this is the crushing twist to the irony—which no creature except man is capable of comprehending. Every creature fights for its own, individual life, refuses, by fighting, to accede to the idea that the salmon or the gnat is of lower order of importance than the idea of the salmon or the idea of the gnat."

Instinctively, I want to rebel against the accusation in the argument, that as a scientist I am somehow culpable, or even complicit. We have spent months with the penguins, learned a little of their histories, given them names, watched them die. All these acts have served to make them more real to us, more vital in our memory. That is what I tell myself, at any rate, but I also understand it is an easy rationale, if not quite a lie. To work in the field is to not dwell on how much we abstract the animals, even as we pull them apart in this way or that so we can fit them into different datasets. It is the nature of exactitude in science. Here on South Beach, the penguin and the southern giant petrel are and always have been data above all. That is the only satisfactory justification for why I am here, am allowed to be here. If this analysis is to have any validity, the details of this death will have to be coded, categorized, smoothed over. Any stubborn bits of individuality will be shorn away—will be called, in fact, error.

The routine of the struggle—if a duet so desperate can be called a routine, if an outcome so assured can be called a struggle—nears its close: the penguin fights to draw breath, the giant petrel fights to deprive it of breath, each lives out its own dreadful present.

I sit on the beach with my hands in my pockets. I could record in the notebook the full measure of what I see. I could turn the penguin and the southern giant petrel into ecology. Instead, I scribble a quick, throwaway line or two. It will be ignored, and that's fine; I suspect it is the point. We already have gathered so much data this year. Better not to be a glutton (or a stinker). Better that this pair should flare bright and then fade into searing afterimage, anonymous and unforgettable.

Even though its strength is spent, the penguin seems to take forever to die. Life, always so reluctant to let loose its hold! I'm not sure how long I stay. Maybe twenty minutes, maybe half an hour, maybe longer, until, at last, the penguin is still. By now, the sun has sunk to the lip of the hills. It illuminates the ocean with its flaming golden light, and this lets me watch in brilliant detail as the giant petrel starts tugging on the penguin's flesh, trying to find purchase, before it gives up and tows the body to shore, where it can more easily tear the thing apart.

A southern giant petrel cruises over the beach.

13

Waiting for Turbo

The first nest, 817C, is empty. In the next, C02B, the second chick basks in the morning sun, but not for much longer, I hope. "When are you going to quit living in your parents' basement?" I ask. The chick closes its eyes, tunes me out. I turn to C03. This nest is also empty; it has been for days. C09E is empty, too, but the female from it, 60510, dozes in a bush a few meters away. Her chick fledged a couple of weeks ago. At C09D, the pair has been gone for a while, as has their surviving chick. In their place sits a male, 53071. He never got a mate and has been shuttling between bushes. "Better luck next year," I say to him. He head-wags at me.

Ten minutes later, El and I have finished the complete morning Cañada check. All the area checks are quick now that the season is all but done. Many of the adults are gone. Save for the chick in C02B and one or two more, all the survivors have fledged. They join tens of thousands of other chicks from Punta Tombo, Cabo Dos Bahias, and colonies elsewhere, all swimming north to the waters off northern Argentina, Uruguay, and Brazil. There they will spend the austral winter. Perhaps 10 percent of them will survive that first year. I try not to dwell on that.

El and I walk back to the house for breakfast. "I'll be in in a sec," I say as we prop our *ganchos* against the wall. El smiles and goes inside.

I trot over to the big bush I know so well, the one with the barbed wire. I kneel down and peer in. Turbo isn't there. He left a couple of weeks ago, and we haven't seen him since. I get up and dust off and trudge back to the house. Tomorrow El and I will leave, and I'm hoping Turbo will come back before we go. Every time someone has left for good this

year—Ginger in October, Emily in December, Briana in early January—
Turbo has returned from an extended absence. We joked that he always
made an effort to say good-bye to the people he liked. It was good for a
yuk then, but now I am anxious. I want to say good-bye to him, want him
to want to say good-bye to me.

⌒

When I go inside, El is making coffee and Dee is sweeping the floor. Dee
doesn't have to teach this term in Seattle, so she came to Argentina for a
few days to help with an extensive coastal survey. She dropped by Punta
Tombo yesterday. In the morning, she will drive El and me to the Trelew
airport. Dee being Dee, she has suggested we do another beach survey
after we finish breakfast. "A last one before you go," she says. "To send you
on your way."

She and I head out to North Beach, clickers in hand. The beach is
dense with penguins, denser than I've ever seen, and for the most part
they look awful. The season has been hard on them, as all seasons must
be. Their plumage is a lusterless brown, and so worn that some birds have
patches of exposed quills on their backs. They need new feathers if they are
to survive the winter at sea. This process of rejuvenation, called the annual
molt, is an energetically expensive and, I'm told, painful procedure. Magellanic penguins undergo what is called a catastrophic molt: they lose and
grow back all their feathers at once. To do this, they first leave the colony
and forage for two weeks or more to fatten up. When they return, they
stand in a state of engorged dishevelment, shedding weight and old plumage, which falls off in clumps. They next replace their down, leaving them
gray and rumpled, like old pilled blankets. Then their dense outer plumage
grows in, little by little. Three weeks later, when the molt is finished, they
are half-starved and their belly skin sags, but they are soft and beautiful,
their backs a rich charcoal against a white belly so bright it seems to glow.

Dee and I make our way through the molting hordes, *click*ing. Loose
feathers blow around our legs like snow, gather in drifts around the
bushes. My eye seeks out the few fresh, radiant penguins that stand here

An adult in the midst of molting its old plumage. First its down and then its new outer plumage will grow in. The entire process takes about nineteen days, during which the penguin is land bound.

and there among the scruffier lot. Dee is drawn to other things. Midway down the beach, she spies a breastbone in the wrack line. She picks it up and turns it in her hands.

"Look," she says. "Predation. A sea lion did this."

"A sea lion?" I say. "How on earth can you tell?"

"See this?" She points to a jagged hole in the middle of the keel. "That's a bite mark. From a canine."

"But how can you tell it's predation from just a bone? Couldn't the sea lion have been scavenging an already-dead penguin or something?"

"Nope," she says. "They don't do that. This is predation."

"Huh," I say. "Wow." Inwardly, I castigate myself. All this time, and I never knew to scrutinize every last bone! Have I learned nothing? Outwardly, I ask, "Should I make a note?"

"Nah," she says. "It's just interesting." She tosses the bone aside, stands and stretches, inhales the breezes, exhales in a rush, exclaims, "Goodness, what a day!"

Surveys weren't always as nice as this. When Dee first did them in the early 1980s, she often walked between two distinct groups of penguins:

the living gathered along the water's edge, and the dead lying where the tide had deposited them in the wrack above. Many of the penguins' bodies were smeared black with oil. For Dee, this was alarming. She knew of no major spills in Argentina, but the carcasses were fresh, so the penguins must be running into oil somewhere.

To find out how widespread the problem was, Dee and some Argentine students began a survey of the Chubut coast in March of 1984. They walked two kilometers at each of fourteen sites, from a beach at Maqueda, near the oil terminal at Comodoro Rivadavia, north to Punta Norte, on the tip of Peninsula Valdés. For eight years they did this, always in March. They counted hundreds of dead penguins, both adults and fledglings. Up to 66 percent of the bodies were oiled. From this, they estimated that more than forty thousand penguins might be dying each year due to oil pollution along the coasts of Chubut and Santa Cruz provinces alone.

Where was the oil coming from? The likely source, they eventually learned, was the ballast water from large ships. To avoid the cost of treating it once docked, tankers would illegally vent their ballast tanks a few miles from shore before coming into port. The penguins swam through shipping lanes rank with oily bilge and were coated. To birds, oil is pernicious. A small dollop might not kill a penguin outright, but it would compromise their feathers' ability to insulate. Chilled penguins were driven to land, unable to forage since to enter the water was to die of hypothermia. Instead, they starved to death on the beaches. When Dee performed necropsies, she found many of the penguins had ugly lesions

Near the end of the season, hordes of molting penguins crowd North Beach as far as the eye can see.

along their stomach walls. In trying to preen the foulness from their feathers, they had ingested it.

The illegal dumping of ballast water presented an enormous challenge, both logistically and conceptually. When people hear "oil pollution," they tend to think of a single spectacular event. A tanker runs aground or sinks, and thousands of gallons of oil spill out of its ruptured hull. People rush to clean the sea, to mop up as much filth as they can, to save pathetic flocks of sopping black seabirds. But when a tanker vents ballast water in the middle of nowhere, no one notices, no one cares. The slick is simply a byproduct of daily life and global commerce, and so is invisible, save for the trail of animals left sickened and dying in the wake.

What was the best way to address chronic oil pollution in a managerial space as fluid as an ocean? Dee and her students somehow had to capture and hold the attention of large, cumbersome institutions, governmental and otherwise. They began by giving public talks on the ways chronic oil pollution affected penguins. They took part in hearings and presented their data as often as they could, to anyone who would listen. The work was slow and patient, but in 1997, thirteen years after their initial surveys, the provincial government of Chubut moved the shipping lanes about twenty-five miles farther offshore, out of the penguins' path. Since 2001, Dee and her Argentine colleagues have repeated the coastal survey every other year. In the four surveys following the lanes' shift, they have found, total, four oiled penguins. It was such a survey that brought her here yesterday. She and Esteban Frere, a former student who now works for BirdLife International, returned to their beaches, walked their two-kilometer stretches. "We didn't see a single oiled bird," she told El and me. "Not a one."

In conservation, success stories can seem rarer than even the rarest animal, but they do exist. "Then we helped get the shipping lanes moved, and the penguins stopped being oiled so much." I had heard Dee say this in classes or public lectures many times before I came to Punta Tombo. Walking with her today on the beach, I understand that I never appreciated the weight and history behind that sentence. I didn't realize the extent to which she took all that time and effort and polished it into a quick spiel, itself just a part of her larger penguin narratives. Perhaps it is because the shipping lanes are done. They are in the past, and Dee is no ruminant. She is quick to say chronic oil pollution is far from being solved in South America, or anywhere, but she has simply turned her focus to different regions. In northern Argentina, Uruguay, and Brazil, toward which the newly fledged chicks are migrating right now, people have rescued nearly four thousand sickened birds over the past twenty years, and an unknown number of dead penguins have washed up on unvisited shores. It is a familiar story; oil will always be a problem where penguins and ships cross paths. Right now, Dee and Popi are assessing the extent of chronic oil pollution along the penguins' migratory corridors. They and other researchers have contacted rehabilitation centers along the Argentine, Uruguayan, and Brazilian coasts, asking them to document what they see. They are collecting data, marshaling evidence, putting together plans of action. I don't know how long it will take, but I am confident that someday this, too, will be a short sentence in the long life of Dee Boersma.

Dee and I finish the survey and turn for home after a couple of hours. Rather than go back along the beach, we walk inland along the berm. We are talking about the past six months, about what El and I have learned, about whatever is on our minds, when another of the countless amusements on the campo catches Dee's eye. She slows to a stop. "Look at that guy," she says. Midway up the berm, a male penguin is eyeing a thick branch. The branch has to be about three feet long. The penguin nibbles it, grips its end, hefts it up, and starts backing his way uphill, tugging the branch behind him.

"That's a heck of a nest decoration," I say. Dee smiles. Penguins, she tells me, can experience a brief upsurge in their reproductive hormones at the close of the breeding season. "They get nesty," she says. Usually, it's a bird that never got a mate, or a pair that lost their eggs or chicks. They might take up residence in an empty bush for a few days. Or, like this fellow, they spiff up a burrow.

We watch the penguin's progress, if it can be called that, as he lugs the branch up the berm. He arrives at last to his burrow and maneuvers the branch to shove it in, but it is twice as long as the entrance is wide. He props the branch across the entrance lengthwise and gives it a mighty shove, trips over it, and loses his hold. The branch tumbles to the bottom of the berm. The penguin gets to his feet and watches it go. Dee can't help herself. She laughs and laughs. "What on earth were you thinking?" she asks. The penguin glances over at us, and waddles down to try again.

⟜

In 1974, the philosopher Thomas Nagel published a celebrated essay titled "What Is It Like to Be a Bat?" He argued that his question was in a basic way unanswerable. We might be able to imagine what it is like to be a bat, but we cannot know what it is like to catch a moth in our mouth, what it is like to hang upside-down in a state of waking sleep, what it is like to perceive the world primarily through sound. Experience is necessarily subjective. When we try to describe a bat's life in ways comprehensible to us—using metaphors, analogies—we push the thing away from itself.

More than once, as I have watched the penguins do their penguin things, I questioned whether it is so hard to know another creature's state of being. "Even without the benefit of philosophical reflection," Nagel writes, "anyone who has spent some time in an enclosed space with an excited bat knows what it is to encounter a fundamentally *alien* form of life." But setting aside our revulsion, is an excited bat really so alien? Presumably it is excited because it is afraid. Surely most of us have felt fear at some point in our lives. Perhaps some of us have even feared for our lives, as the bat does. We know what it is like to be *that* bat, if not *a* bat.

To suppress disgust and identify with that bat's fear shows empathy, or compassion, which for those of a certain bent might lead to a wish for greater understanding. Because someone must have been fascinated by a frantic bat flying around an enclosed space, we now know more about what that bat was doing. We know a bat can generate ultrasonic *clicks* with its larynx, sometimes emitting them through its nose, at rates as high as two hundred *clicks* per second. We know that when a *click* bounces off an object—a moth, a wall, my head—it returns to the bat's elaborately structured ears, themselves slightly misaligned on the skull. From there, the *click* travels to a specialized inner ear, where it hits a disproportionately long and thick basilar membrane within the cochlea, stimulating special nerve cells. These signal the highly developed inferior colliculus and auditory cortex in the bat's brain. We know that in response to these impulses and signals, which take place in milliseconds, a bat can twist and drop and dive through an entirely dark space with extraordinary precision.

We know all of this because some scientists spent more than thirty years teasing out these mechanisms, which might lead to another question: What is it like to want to know what it is like to be a bat? In a speech he gave in 1918, Albert Einstein suggested an answer when he observed that scientists can be marked by a peculiar "state of feeling." This feeling, he said, was "akin to that of a religious worshiper or the lover. The daily effort comes from no deliberate intention or program, but straight from the heart."

Belief. Love. These words are not often heard in the sciences, but strong feelings for one's subject might not be so rare, especially in biology. (Outside of its philosophical implications, "What is it like to be a bat?" is, at heart, a biological question.) Why else would someone study an animal for thirty years, for forty years, if not for a kind of love? Maybe someday I will ask Dee. I imagine her answer will be as obvious as it is professionally delicate.

Love is a feeling not lightly given. Nagel was careful to ask his question using an animal with which people might identify. He chose bats, he said, because humans are closely enough related to them to grant that they have conscious experiences. Head too far down the phylogenetic tree, he wrote, and "people gradually shed their faith that there is experience there at all."

Were he willing to move from Mammalia to Aves, I think Nagel could just as easily have asked his question about penguins. Excepting the great

apes, what animal is more like a human than a penguin? The kinship
has been felt for as long as penguins have been known. "They walke so
upright," wrote John Winter, who sailed with Drake around Tierra del
Fuego in 1578, "that a farre off man would take them to be little children."
Nearly four hundred and fifty years later, little has changed. "People can
identify with penguins," Dee once said in an interview. "These birds are
curious. They walk upright. They dress well. They're highly social. They
know their neighbors."

Perhaps, then, in a nod of acknowledgment to our shared fellowship
as animals, we can know in some way what it is like to be a penguin.
Likewise, penguins might have a sense of what it is like to be human,
squabbling as they do among themselves, trying to find a mate, struggling
to raise a brood against daunting odds. But now I have gone too far. I am
anthropomorphizing. Undoubtedly, Nagel is right. Despite our pokings
and proddings, our stunts and interventions, it is impossible to know
what it is like to be a bat, to be a penguin—to know what it is like to be
something we are not. Yet that has not stopped Dee from trying.

⌒

Dee and I get back to the house at lunchtime. I glance over at a certain
large bush.

"Waiting for Turbo?" she asks.

"Yeah."

"Still not back yet?"

"No."

"He's probably out foraging for the molt," she says. "I bet he comes
back."

I hope she's right.

Anthropomorphism means, basically, "human form." The word was
first used by Xenophanes, I have read, a philosopher and poet who
objected to the way Homer described gods as though they looked like
humans. Xenophanes wanted none of this. If horses had hands, he is said
to have argued, they would draw their gods like horses.

Sometimes I wonder how we would draw our gods if we had flippers like a penguin. Or I wonder how Turbo would draw his gods if he had hands like us. But he is not here for me to ask.

After lunch, El and I change out of our work clothes and boots and pack them away. Some of the stuff we will bring to Seattle, but most we will box up and store in the *cueva*, should someone need them next season. Sartorially liberated, we walk down the tourist trail. It feels nice to play tourist when so few are here. We pass through an empty campo and look out at a calm sea. I say something about how this is what Punta Tombo will be like for the next few months, but catch myself. This is not what Punta Tombo will look like, for some penguins are still lying outside their nests. I'm not sure what keeps these holdouts here, or why they are reluctant to leave, but I know the feeling.

We reach the end of the trail and turn and come back. When we are near the house, again I break off and jog over to Turbo's bush. It is a reflex. He isn't there.

"He'll come back or he won't," El says. "Checking all the time won't change that."

True. It is hard for me to articulate why my vigil has grown so desperate. Vanity, maybe, or a need for interspecific acceptance. It takes no special virtue to love a penguin that loves you back, but I have still shown something of myself to him. I may not be of this place, but I wish I could belong, and Turbo is the conduit.

"I'll feel like a bad person if he doesn't come back for us," I say finally.

A few days ago, I saw a big male penguin disappear into Turbo's bush, but when I ran out and called Turbo's name, the penguin peered out blankly and head-wagged. Not Me. I sat down in the dirt and moped. The penguin inched out, head-wagging all the while. When he was clear of the bush, he turned and ran from me. As if I were the intruder, the crazy one.

Now it is evening. The house is bare, everything packed in the *cueva* under lock and key. Our bags are piled by the door. Dee has gone to bed. El and I sit at the table.

"Are you glad we did this?" I ask, not for the first time, but probably for the last.

"Yes," she says. "Yes, of course I am."

"What will you miss?"

El thinks for a moment. "I'll miss the penguins," she says. "And the chicks." She sighs. "I miss the chicks."

"Would you want to do it again?"

El smiles a very El smile and gets up to go to the trailer. "No, I don't think so," she says. "Once was enough."

Now it is night, and I drink the dregs of hope. Dee drives us to the airport in a few hours, and then we will fly away, and we will never come back. What are the chances Turbo returns before then?

"I'm going to go check his nest," I say.

"You know it's okay if you don't see him again," El says. "It doesn't mean he doesn't like you."

"I know," I say, feeling foolish. "I just want to check."

"I hope he's back," El says.

I hop out of the trailer and approach the dark shadow that is Turbo's bush. "Turbo?" I call. Before the sound of his name has even faded, Turbo erupts from his nest, braying wildly. He is hugely fat and still wet from the sea. A piece of red algae is stuck to his band.

Later, El will tell me she thought someone had run over a dog, so demented are my happy howls.

Turbo and I go through our ritual of flipper-patting, and then I accompany him back to his bush. He scoots past the branches and ducks under the barbed wire, levering his bulk with his bill before slipping away into the dark. I can hear him clucking, which is what penguins do when they are home. I stand for a moment outside his nest so I can listen to him go about his domestic routine, feeling relieved and in a way justified. Then I realize Turbo brayed as he would have had he been greeting his mate after a long absence. I am not sure what this says about either him or me, or us. I decide not to worry about it too much.

I start back to the trailer. The sky is black, with a bare sliver of moon. Thousands of penguins are braying all around me, and I am subsumed in the symphony of their din. How curious it is that their display is called "ecstatic." Ecstasy is an emotion we humans typically reserve for ourselves. Animals, when we give them credit for feeling anything at all, are more like to be afraid, enraged, jealous, concupiscent, or simply blank. We are less sure what to do with acts that look surprisingly, suspiciously like joy. An amusing joy, true: a braying penguin is a comic sight. But when heard at night in chorus with a quarter of a million other birds, the bray becomes something else. It has a strange, haunting power, as all the ecstatic displays merge with one another into a great wall of sound, and the fermata of the *hooAAAAAAAAAAH* rolls out over a throbbing ostinato of *hu-hu-hus*.

When we leave the penguins the next morning at dawn, the desert still rings with the ecstasy of their singing.

Penguins bustle about in the evening.

The Bleak Land

> In calling up images of the past, I find the plains of Patagonia
> frequently cross before my eyes: yet these plains are pronounced
> by all most wretched and useless. They are characterized only by
> negative possessions; without habitations, without water, without
> trees, without mountains, they support merely a few dwarf plants.
> Why then, and the case is not peculiar to myself, have these arid
> wastes taken so firm possession of the memory?
>
> Charles Darwin, The Voyage of the Beagle

The big plaster penguin and its chick are gone now from the Marco
A. Zar airport. In their place is a softly lit wall panel with pretty calendar
shots of Argentine Patagonia. I don't see a penguin until I pass an ad in
arrivals for a company called LondonSupply. The ad reads, *Los mejores
destinos merecen los mejores aeropuertos* ("The best destinations deserve
the best airports"). Beneath the text are photos of a penguin, a glacier,
and a lighthouse, meant to exemplify, respectively, the cities of Trelew, El
Calafate, and Ushuaia. But the penguin isn't a Magellanic penguin. It's an
African penguin, a close relative, yes, but one that lacks the collar across
the throat, having only the inverted U on its breast. How on earth could
LondonSupply get the species wrong, and here of all places? Don't *Los
mejores aeropuertos merecen la especie de pingüino correcta*? I guess not.

I haul my bags to the main terminal, where I see a bigger version of
the blasphemous LondonSupply sign, again with the African penguin.
No, wait—on this sign, someone has scribbled in the missing collar with

a black marker, turning the penguin into a proper Magellanic. Who was the vandal responsible? Dee Boersma, of course. As if you had to ask.

⌐⌐

It is November of 2015. I finished my PhD in biology a few years ago but have not stayed in science, choosing to tell its stories instead. The penguins stayed with me after all this time, though, and so I talked Dee into sending me back to Argentina, if only for a few weeks. El stayed behind in Seattle with our young daughter. "There's no way I could handle the chicks now, as a parent," she said while driving me to the airport. "But say hi to Turbo for me."

Soon I am sailing out of Trelew in the backseat of Opus with Caroline Cappello, Katie Ferguson, and Liz Muntean, who make up this year's crew. The closer we get to Punta Tombo, the more it might be seven years ago, when El and I were seeing this place for the first time, or seventy, or however old Ruta 1 is. The sky and the land are the same, wonderfully the same, and Opus flies down the highway, flinging stones as ever.

An hour later, Caroline turns down the last road to the colony, and it is then that my gut gives its first good lurch. "Just wait until you get there and see how much has changed," Dee had told me before I left, so I had some warning, but it is another thing to see all the changes for myself. Foremost among them is that the crew no longer lives in the old field house. The *guardafaunas* took it over after Dee was denied her research permits one year and had to cancel an entire season. She didn't get the house back the next year, but the provincial government has since built three small houses half a mile or so inland, along with a café and a museum. For now, the crew stays in one of the new houses, a rambler with huge picture windows that look out on the campo, with the Atlantic some distance beyond. Although it is a much more habitable space, I can't get over how far we are from the colony. The distance is a spiritual barrier. I step outside and hear nothing but the wind, or laundry snapping on the line.

After I've stowed my luggage, I join Caroline on her evening check of the birds currently carrying satellite and, as of 2012, GPS tags. We drive down to the park entrance and walk into the colony. I wait for all the old sensations to flood over me: the sights (penguins everywhere), the smells (fishy guano), the sounds (*hooAAAAAAAAAAAH!*). But there is a blurriness, as if my senses have dulled. The penguins seem smaller. I have trouble telling males and females apart. Reading bands is hard, too. Were the numbers always so small?

I follow Caroline as she strolls through the campo, twirling her *gancho* around like a vaudevillian's cane. The tagged birds are clustered in a place Dee calls RFID Land. A new area, RFID Land combines nests from White Flats, Mesa, and Hill-Draw—older areas the crew doesn't check regularly anymore. What makes RFID Land special is that all its study penguins have, in addition to the usual flipper bands, radio-frequency identification tags, or RFIDs. The tags, which are the size of a rice grain, are injected into the penguins' left feet. Across a narrow *cañada* the penguins must pass through on their way to the ocean, Dee has spread three scales with RFID readers in them. The penguins are funneled over the scales one by one, and each gets weighed. Any with an RFID in its foot is recorded in a computer. The scales can tell the direction the penguins are heading, so in addition to the times they come and go, Dee knows how much they weigh when they leave the colony, how long they are gone, and how much they weigh when they return. The system has spawned a new series of questions for her: about the relationship between foraging success and the openings and closings of nearby fisheries; about the habits of successful versus unsuccessful breeders; about annual differences in food availability and foraging effort. "Now they're never going to get rid of me," she says.

Caroline goes up to the first nest, a burrow. "*Permisssso*," she says sweetly, nudging the male penguin with her *gancho* so she can read his band. I see a small GPS transmitter about the size of a Zippo lighter glued to his back. He has one enormous chick, and an egg that will probably never hatch. He head-wags at us.

Caroline *permisssssos* her way through the check, and we turn back. "So, yeah, that's what it's like now," she says. "How different is it from when you were here?"

"You have no idea," I say. I could go on and on, but I'm aware that, if I'm not careful, every sentence out of my mouth will start with, "When *I* was here . . ." I'd hate that as both speaker and listener, so I look at the penguins instead. They are as timeless and ageless as ever. I walk past the unbanded occupants of areas I used to know, and I wonder if they were alive when I was last here, wonder if, by some small chance, they remember me in a dim way. Of course, when I wonder all these things, I have only one penguin in mind.

⌇

"He has his same bush," Caroline says, and she is adding something as I skip past her, so I don't catch it. "Hey! Turbo!" I call. The penguin at the heart of the big *molle* bush with the barbed wire head-wags imperiously. "But he's on the backside now," Caroline finishes. Oh. I walk around until I find Turbo in a dusty bowl on the bush's rear margin. This new nest has no cover other than a couple of twigs. Turbo is resting on his belly, his back to the world.

"Hey, Turbo," I say, more hesitantly this time. He gives no indication of having heard. He hasn't preened himself in a while, and his plumage is dusty and ratty, with stray tufts of down poking out. Someone, I see, has pooped on his back. Worse, he is incubating a small black rock. He glances up at me, adjusts himself over his rock, and spreads his flippers protectively. His head trembles when he does this. How do I put it nicely? He looks like he has lost his mind. I think of the mad king Nebuchadnezzar wandering in the wilderness, naked and snarling, with his featherlike hair and bird claw fingernails.

I squat down and put my hand on Turbo's back. He doesn't seem to notice. His head still trembles. "Um, we're not supposed to interact with him," Caroline says gently. The *guardafaunas* have really cracked down this year. No one is to touch Turbo when he approaches, or to be friendly with him, ever, regardless of whether or not tourists are here. Turbo is wild, and to be treated as such.

Turbo incubating his rock when the author first arrived back at Punta Tombo.

Over the next few days, I reacquaint myself with the colony. The penguins started breeding late this year—snow was still on the ground in September—and so the chicks are just now beginning to hatch *en masse*. Between them, the scales in RFID Land, and sundry other duties, the days are full. During the first days, I'm not sure how I will manage. When I go to the *cueva* to fill my vest with gear, I can't recall where I put everything before, so I stick stuff in pockets willy-nilly. Then I go to weigh a penguin, and my right hand drifts automatically to the vest's left front pocket. Oh, yes, that's where the scale went! Thus am I reassembled. Still, I can't shake the image of Turbo and his rock. Caroline and Katie are taken aback at my hand-wringing, and Liz assures me Turbo is fine. When Dee was here a few weeks ago, he was his usual dapper self, and she happily ignored the dictums of the *guardafaunas* to pet him.

I am not reassured. Every morning, we drive the quarter-mile down to the colony for the Cañada check. Every morning, while the others note the weather, I duck around the bush to call on Turbo. Every morning, I find him sitting on his rock, dusty and trembling. He has become one of those sad single males that incubates garbage.

This goes on until one morning I walk around the bush and Turbo is gone. He has left to forage. Thank goodness. At least he can still take care of himself, and had enough sense to leave the rock behind. That damn rock. There it sits in the dirt, black and vaguely sinister. I stare at it, and confront a sudden and powerful dilemma.

Argument: Take the rock. Set Turbo free.

Counterargument: Leave the rock. Turbo doesn't need my help and never did. He manages as best he can, as do all the penguins. One rock won't change his fate or theirs. In truth, I would be taking the rock for myself, not for him, and I shouldn't involve him in the amateur dramatics of my sentimental theater.

Turbo's rock, after he left to forage.

I turn on my heel and hurry away, leaving the rock where it is, but I can't stop obsessing over it. It becomes a totem: of distraction, of wasted effort, of needless worry, of the denial of your true nature, of trying to be something you're not, of making claims on something that isn't yours, of who knows what else. Finally, I can't stand it anymore. When Turbo hasn't returned in a couple of days, I go to his nest and take the rock and put it in my pocket. I'm not proud of this, but I can't help it, either. In the days after, when I'm walking idly about the campo, my hand often strays to the rock. It fits nicely in my palm.

⌒

Having arrived just as the chicks are arriving, I find that while I had forgotten how fragile and soft they can be in the hand, I will never forget how easy it is for them to die. Die they do, but if I feel the old gloom whenever I reach for my egg cup to remove a wasted body out from under a head-wagging parent, I sit in a lighter shade of despondency when I weigh and measure the remains. I suppose I am better reconciled to my helplessness; or maybe it is more honest to say I am inured to it.

During one morning Cañada check, I watch Liz scoop handfuls of dust over a chick that succumbed to starvation the night before. As we sit in silence next to the little burial mound, I remember how, years ago and in similar circumstances, I had gone out of my way to avoid using the word "bleak" to describe this part of Patagonia. But how is this land not bleak? The question comes to me again and again as I move around Punta Tombo, trying to make sense of its difficult beauty. Dee had said much has changed, and that is certainly true, but most of the changes seem cosmetic to me. A few houses, a museum that almost no one visits so far as I can tell, the café, some additional boardwalks, some wire fences, a couple of signs. These are insults to my memory, but not much more than that. The great things, the things that matter—the altered climate, the loss of habitat, pollution, the number of fish in the sea, the pattern of their movements, our appetite for them—are much the same for the penguins, if not worse.

And yet. A couple of mornings after Turbo has left, we get word that, at the last moment before a new administration enters office, the Chubut government has passed legislation to establish the Punta Tombo Marine Protected Area around the colony. The area will cover thirty-seven miles of the Chubut coastline and extend three nautical miles off the coast. Within that zone, fishing will be curtailed during the penguins' breeding season. The legislation also calls for conservation measures that will apply up to twelve nautical miles off the coast, although it isn't clear what exactly that means in terms of actual enforcement. Regardless, this will be only the second protected area in all of Argentina that is exclusively marine.

Publicly, Dee and Popi both talked of the marine protected area as a tremendous victory, one for which they had vigorously campaigned for a long time. "We've been protecting the penguins on land for thirty-three years," Dee said in a press release. "Now they finally have their food protected at sea." Popi especially had worked hard. He drafted the proposal to design the protected area, and he traveled back and forth between Rawson, the provincial capital, and Puerto Madryn, where he lives and works. The legislators needed regular prodding, and he provided it. "This wouldn't have happened without him," Dee said.

All of this followed the creation of Patagonia Azul, a new UNESCO Biosphere Reserve that stretches from just north of Punta Tombo all the way down below Cabo Dos Bahias. In all, that reserve encompasses not only the protected area, but also more than three million hectares of land and sea. Its aim is to encourage sustainable development and responsible ecotourism, preserving the region's unique character. "Two thousand and fifteen has been a great year for the penguins of Punta Tombo," Dee said.

I think again of what *bleak* meant to El and me before and weigh that against what it means to me now, in these times. During the off hours, I have been reading a book of old radio interviews between Jorge Luis Borges, the great Argentine writer, and Osvaldo Ferrari, a young poet. In one exchange, Borges observes that the English words *black* and *bleak* share the same root as the French word *blanc,* the Italian *bianca,* the Portuguese *branca,* and the Spanish *blanco,* each of which means *white.*

BORGES: It's odd that this word branched out into two opposite meanings—we tend to see white as opposed to black. But the word it proceeds from means 'without color.' So, as I've said, in English it suggested the dark, meaning black, and in Spanish clarity, meaning white.

FERRARI: There's chiaroscuro in etymology.

BORGES: That's right, chiaroscuro, an excellent observation.

Black and white. Darkness and clarity. When I am back in Seattle and visit with Dee at her office, she will tell me the protected area is just a postage stamp of what she and Popi had wanted. Nonetheless it is an important start. "That's how conservation works," she will say. "You move little by little, you keep pushing, and eventually something gets done." So yes, Punta Tombo will always be bleak. Bleak in every sense of the word.

〜

A few days later, out of habit more than need, I walk to the rear of the barbed wire bush and there is Turbo, back again from the sea. He is fat and clean and looks so much better, confirming, if I ever doubted it, that he will always be an exceptional penguin. I have to laugh, too, because of course he has found another rock to incubate, and this one is even bigger than the first. I pluck it out from under him—it is positively roasting from his body heat—and then give it back. He tucks it under himself and eyes me in that penguin way of his. *There,* he seems to say. *Now we both have one.*

I kneel beside him, make sure no one is watching, decide I don't care even if they are, and run my hands over his back, scratch his neck, knead his feathers. He clucks, closes his eyes, and wriggles with what I can only hope is contentment.

〜

Turbo, having returned from foraging, incubating a new rock.

The closer my departure looms, the more I try to slow the days. Some-times in the late afternoon, I stand atop the berm and watch the penguins down on the beach, bathing, calling together, resting on their bellies. In this state of ecological beatitude, with the sun sinking and the sea lumi-nous, I try to focus on each second as it *ticks* remorselessly by, all the bet-ter to hold on to the aching moment when lived experience and memory are one. But it is not to be. The wind blows, time marches on as ever, and then it is my last morning here.

The day is cool with a heavy mantle of cloud, but the unseen sun blasts colors across the available slab of lower sky. Caroline, Katie, Liz, and I are checking the Factura area. Once we are done, we will all drive to the airport in Trelew. The crew will stay in town until the evening, when Dee flies in. She and I will just miss each other.

The prospect of not making my flight puts an end, mostly, to any spiritual dilly-dallying. Whatever I have now is what I will take with me, and so I tromp briskly about the campo, attending to the living and the dead. When I'm through with my allotment of nests, I go find Katie to see if she needs a hand with anything. She is in the midst of weighing a pair of large and unruly chicks, and nods to their burrow. "The female needs to be toe-tagged," she says. "Do you want to do it?"

Oh, do I. I reach into the burrow with my *gancho,* pull the female out, and grab her by the neck. Weigh her first. She is small and light, only 3.5 kilograms, or 7.7 lbs., but otherwise healthy. Calm, too. I barely need to hold her while I measure her bill's length and depth, then her flipper, then her foot. I put my head near hers and loosen my grip until my fingers are just lightly against her. Her eyes are dark, as are her feet. For some reason she strikes me as old. I will think of her as old.

After I finish noting her particulars, I pull a toe-tag out from the film canister, record the number—10874—and stick it through the webbing on her left foot. I hold her for a moment more, tt10874, trying to memorize the details of her body: the feel of the soft but scaly texture of her feathers, the strength of her neck, the ridges and furrows on her bill, the hard curve of her flipper, her mottled feet. I would remember everything

A female and her two chicks in their burrow.

if I could. I would split myself in two to be both here and home. Then she twitches in my hand, jarring me from my reverie. Time to let her go. I point her toward her burrow and release her. She waddles back to her brood. They nuzzle each other.

Acknowledgments

To start with the most obvious and necessary, thanks to Dee Boersma for being willing to send El and me to Punta Tombo all those years ago. She promised the experience would be life-changing, and it certainly was, although perhaps not in the way she necessarily envisioned or hoped. Briana Abrahms and Emily Wilson (now Emily Connelly) were wonderful coworkers for the months they were there. Likewise Jeff Smith, Olivia Woods, Clay Gravel, Julian Nuñez, and Stephanie Mitchell, all of whom came to the colony for different periods of time. Popi Garcia-Borborglu was always a source of good cheer whenever he dropped by. Thanks to the La Regina family and the Provincial Reserve for allowing us access to their lands, and the Wildlife Conservation Society for financial and logistical support. Thanks to Caroline Cappello, Katie Ferguson, Liz Muntean, and Thaís Stor for making me feel more than welcome when I went back to Argentina in 2015. (And, again, thanks to Dee for letting me return.)

I've been indebted to many people throughout this project, even before it was a book. Laura Helmuth, then the science editor at *Smithsonian*, commissioned a series of dispatches back in 2009 that helped me approach the experience with narrative in mind. I'm especially grateful to Chip Blake and Hannah Fries of *Orion*, who in 2011 published "I, Turbo," the article from which this book grew (or hatched). Other portions of the story also appeared in *Audubon* in a substantially different form. More recently, I'm especially especially grateful to Ginger Rebstock for reading the manuscript almost from start to finish, and saving me from as many embarrassing mistakes here as she did when she was teaching El and me how to make our way through the field. Any errors that remain are completely my fault, which should go without saying.

Mary Braun, Marty Brown, Micki Reaman, and the rest of the staff at Oregon State University Press have been so kind and helpful from beginning to end. Thanks to two anonymous reviewers for their helpful comments and suggestions. I'm grateful also to Bob Pyle not only for the title suggestion, but also for thoughts and encouragement over the years. Andrew McNair kindly read a portion of the manuscript to give a sense of how it seemed to un-penguined eyes.

Last but not least, thanks to all my family for their love and support. Thanks to Bay, because how could I not? And to El, thanks for everything and more. Sometimes a crumb falls / From the tables of joy . . .

Index